BRAIN POWER

BRAIN POWER

Optimize Your Mental Skills and Performance, Improve Your Memory, and Sharpen Your Mind

Tony Buzan

Published 2024 by Gildan Media LLC
aka G&D Media
www.GandDmedia.com

Front cover design by David Rheinhardt of Pyrographx

Designed by Meghan Day Healey of Story Horse, LLC.

Library of Congress Cataloging-in-Publication Data is available upon request

ISBN: 978-1-7225-0636-0

10 9 8 7 6 5 4 3 2 1

Contents

Preface

Welcome to *Brain Power: Optimize Your Mental Skills and Performance, Improve Your Memory, and Sharpen Your Mind.*

Congratulations. You're about to learn exciting new techniques that will help you remember more, think more clearly and creatively, develop problem-solving and analytical skills, read and study with speed and efficiency, and climb the ladder to business and personal success!

You'll be guided by Tony Buzan, one of the world's most highly acclaimed authorities on learning, memory, and techniques for optimizing brain power. Tony was a captivating lecturer and seminar leader who, over the course of his life, trained thousands of employees at Fortune 500 companies all around the world.

Brain Power is a unique collection of information, instruction, and exercises developed to help anyone with the desire and persistence to improve their memory and sharpen their mind. Relevant topics are presented first to increase your understanding of key concepts: for example, the difference in function between the left and right sides of the brain, why you remember some information better than others, and what Mind Maps are and how to use them. Exercises are then provided to reinforce your understanding of the key concepts and help you master each brain power skill.

Your brain is capable of infinitely more complex accomplishments than you may ever have imagined.

Brain Power will help you unleash your extraordinary latent mental capacity.

Introduction

Wake up. You're about to learn how to use your brain to achieve complete control of your career and personal life. Soon you'll be able to use your eyes to see things you never knew were right in front of you. You'll use your ears to pick up key information that others miss completely. You'll remember the facts and figures that can move you to the top of your field, and you'll dramatically improve your ability to reason, imagine, create, and apply logic. You'll be able to process and manage information in an age when the people who do it best are the people who succeed in business and in life.

Welcome to *Brain Power: Optimize Your Mental Skills and Performance, Improve Your Memory, and Sharpen Your Mind*—a comprehensive program that gives you the opportunity to benefit from incredible advances in the study of brain health and function that are enabling people around the world to more fully tap their latent brain power. The information and guidance in this book will change the way you think about yourself and the world around you. It will also change the perception that others have of you. Your colleagues will see a you they never knew, providing you in turn with opportunities you never knew existed.

How is this going to happen? What kind of magic is going to take place in your mind to effect this miraculous transformation? What will you be like? What will your life be like when you master these brain optimization techniques? Let's take a quick look at what you have to look forward to.

I have been fortunate enough to be able to help hundreds of thousands of people improve their brain power dramatically (and I do mean *dramatically*). In recent decades, there have been so many advances in what we know about the brain and how it functions that we can now help people wake from their current sleeping state (and I do mean *wake*). As you master the techniques I present, regardless of how efficient you think your mental performance is now, you will feel as if you've been asleep for your entire life till now and that you have just at this moment awakened. In exchange for a modest investment of your time and effort, you will receive the gift of a lifetime: a new self-awareness and confidence, an ability to think your way to the top.

Does that sound like a lot to swallow? Do my bold claims have you wondering what I can possibly do for you in this book that you haven't already tried? Think about this before you go any further: Mortimer Adler of the *Encyclopedia Britannica* said that I possess "wide knowledge of the most recent advances in our investigation of the brain's potentialities and of the mental processes that they underlie. On the basis of this knowledge, his materials on the improvement of our memory, our ability to listen perceptively, our skill in reading and in note taking are filled with ingenious and concretely practical suggestions." Adler went on to say that my work on logic and analysis provides a highly useful guide for anyone who wants to make the most of his or her mind. Remember this as you learn and practice the techniques in this book.

Many of the largest and most successful companies in the world have used my techniques in their management and training functions, including IBM, Barclays International, General Motors, Shell Oil, DuPont, Johnson and Johnson, ICI, Rank Xerox, British Petroleum, Flora Daniels, and Digital Equipment, to name a few. In fact, these companies have paid tens and sometimes hundreds of thousands of dollars each to have me teach my information management techniques to their employees. With this book, you have access to the same successful processes at a fraction of the cost. Just give me a few hours of your time, and I can change the way you manage information and your life through brain power.

Here's a statistic for you: The average business executive has spent anywhere from a thousand to tens of thousands of hours in the formal learning of such subjects as economics, history, languages, literature, mathematics, political science, and other subjects. But that same executive typically spends zero to 10 hours formally learning about how he or she assimilates, processes, and retains information and how his or her thinking patterns dramatically affect the ability to cope with change. With the skills you'll develop while reading this book, you'll not only understand how the brain works to attain mental literacy, but you'll put that understanding to work to earn yourself and your company significantly more dollars. You'll improve your reading, writing, listening, and communication skills. You'll develop better relationships professionally and personally, and you'll greatly increase your brainstorming, problem-solving, and decision-making abilities. My techniques will take you to a new level of brain development and performance.

Each chapter of this book is organized for maximum effectiveness and efficiency. The first part of each chapter is informative, bringing you up to speed on a topic or concept. The second part leads you through a series of exercises that enable you to put what you've learned into practice. Working through these exercises reinforces what you've learned in the reading while training your brain in techniques for optimizing memory and cognition. As you work through the exercises, you strengthen your neurons and develop new neural networks—connections between neurons.

Busting the 10 Percent Myth

A commonly held myth is that most humans use only about 10 percent (or some other small percentage) of their brains. The origin of this misconception has been traced, by some accounts, to physicist Albert Einstein, and by others to philosopher William James, who wrote, "We are making use of only a small part of our possible mental and physical resources." Perhaps this statement refers to the fact that we use a small part of our brains for intellectual pursuits.

The fact is that no part of our brain lies dormant, waiting to be put to good use. The entire brain is engaged in numerous tasks throughout the day, ranging from coordinating physiological activities, such as heartbeat, respiration, and digestion, to processing sensory input and directing physical activity. Rest assured, each of us has a brain that is fully engaged, with a virtually unlimited capacity to learn.

Building brain power is about making the brain more efficient and effective. Even a small increase in brain power can result in an impressive performance boost, so there is no need to try to hit the ball out of the park on the first pitch. Think about building brain power as an incremental process. Don't concern yourself about seeing dramatic improvement by the end of the first chapter. Take small steps, and process the information presented in each chapter in small chunks. Read through the book once, work through the exercises, and repeat the process to reinforce your learning. This program will have a cumulative effect on you. Each skill you learn builds on the previous one, and when you get the whole mix, you're going to find that you continue to improve every day for the rest of your life.

Challenge Your Self-Limitations

Now you may be thinking, "But there are some things I'm just not very good at—math, for example."

It's very important to understand that everyone, and I mean *everyone*, can increase his or her mental ability far beyond its current level. While it's true that genetics plays a role in establishing intellectual ability, the way each of us views our own intelligence is far more limiting. If you place limits on yourself, you won't perform to the best of your ability. You may have had a teacher tell you in second grade that you were hopeless in math, and from that moment on, you didn't even come close to achieving your potential. You are what you think you are. For the most part, your limitations are self-imposed—the product of defeatist thinking. The first step in any self-improvement program is to acknowledge that improvement is possible. Furthermore, research has shown that once you start to improve in math, or any other area for that matter, all other areas of mental performance improve.

At some point in your educational career, you may also have been taught that you are more right- brained (artistic, creative) than left-brained (logical, mathematical) or vice versa. That may be. There really are two brains—a right and a left—and they operate differently. The right side is responsible for your sense of rhythm, images, imagination, color, dimensionality, and daydreaming. The left side deals with words, logic, numeracy, sequence, linearity, and ability to analyze. But there's absolutely no reason why one has to be dominant over the other. Each of us has a fully capable left *and* right brain. Each of us has the capacity to be *both* right- and left-brained.

Probably the greatest example of someone in history with a fully developed and well-balanced brain was a genius who excelled in a wide range of intellectual and creative disciplines, including art, sculpture, physiology, innovation, meteorology, geology, engineering, aviation, architecture, mechanics, anatomy, physics, and general science. Guess who I am referring to.

If you guessed Leonardo DaVinci, you are correct.

On the other hand, many of our greatest scientists and artists appear to have been quite lopsided in terms of brain activity. Einstein and many other scientists appeared predominantly left-brain oriented, while the artists Cezanne and Picasso were right-brain dominant. Einstein failed French in school, but he had a wide variety of interests that no doubt challenged his brain, including playing the violin, sailing, reading, art, and daydreaming.

Your brain need not be lopsided. You can develop both halves, and this book encourages you and leads you through the process of fully engaging both sides of your brain for optimal function. After all, the balanced brain functions best. The most brilliant mathematicians, for example, don't laboriously do the math in their left brains; they visualize problems and solve them creatively by engaging their right brains as well.

Your IQ Is Not the Sole Measure of Your Intelligence

Results of intelligent quotient (IQ) tests are often presented as a sign of a person's raw intellectual ability. For example, they are often used to compare presidents of the United States and other world leaders. Many people have been led by their IQ scores to believe that they have only average intelligence and that they are poorly equipped to take on certain intellectual challenges.

However, research shows that IQ tests are not entirely reliable measures of intelligence. They measure untrained and undeveloped human performance, period. One Berkeley study on creativity showed, for example, that a person measured to have a high IQ was not necessarily an independent thinker or doer, had a sense of humor, or even valued one . . . or appreciated beauty, or originality, or novelty. For that matter, the high-IQ person may not even be knowledgeable, fluent, flexible, or astute. Intelligence tests measure the natural dimensions of our minds, our *capacity* for learning and understanding, but each of us has a nearly unlimited capacity to learn. What separates the superior mind from the mediocre one is how that mind is used and challenged. Just as we become stronger physically when we challenge our body's endurance and strength, the mind becomes stronger when frequently challenged.

Were IQ tests developed to discriminate?

Contrary to what many people believe, the IQ test was not developed to discriminate against disadvantaged students. Absolutely to the contrary: the noted French psychologist Alfred Binet felt it unfair that only upper-class children had the opportunity for higher education. For that reason, he devised the first IQ test—to allow any child with developmental abilities to continue his or her studies. IQ tests, in other words, were the ticket *out of* deprivation; they provided a path to overcoming discrimination. They opened never dreamed of opportunities for children who would not otherwise have had a chance to excel beyond their social class.

My Inspiration

Many people ask what triggered my interest in brain research and the relationship between intelligence and performance in academics, career, business, and life in general. My inspiration came from a childhood experience.

When I was a young boy, my best friend and I loved the outdoors. Everything in the nearby fields, woods, and streams was a new and wonderful experience for us. We both loved nature with a child's unbridled passion. By the time we reached the age of seven, schools had already intruded on our young lives, but we continued to spend as much time as possible out of doors, especially after school let out.

Our lives changed dramatically, however, when our teacher gave us our first test. He asked questions such as, "Describe the difference between a butterfly and a moth" and "Name three types of fish you can find in local streams." My friend who could name virtually any bird or insect, butterfly, or moth simply by observing its flight patterns, received a zero on that test. I, on the other hand, who could not even come close to duplicating his brilliant feats of recognition, received a perfect 100.

From these test results, I was considered intelligent and sent to the top of my class. My friend was considered dumb, classified as unintelligent, and sent to the bottom of the class. It was horrible in a way, because we both knew that the situation was really reversed, but the school didn't. I didn't understand until years later what really had happened, although I was intuitively aware of it. It turned out, of course, that my friend was illiterate and innumerate (he couldn't count): he came from a very poor home.

The only problem—if you want to call it that—was that he hadn't developed a number of basic abilities handled by the left cortex. Because of that, he couldn't handle the test. I, on the other hand, had developed those abilities and could handle a test, even though his brain could compute much more than I could about the subject in which I was considered bright and he wasn't. He was the real genius, just not on paper.

As I continued my studies, I came to understand that many such problems were universal, and they could all be traced back to either a lack of awareness or a lack of training in simple, basic brain skills. I was hearing the same problems from everyone I spoke to: *I can't concentrate. I feel lousy. I'm not happy, but I don't know why. I don't know where I'm going or what I want. I'm afraid of failure. I never seem to have time for anything. My personal relationships aren't what they should be. I'm stressed out.*

Using myself as a guinea pig, I soon found that by applying new findings about brain functions, I could become a master of self-management. I was able to do much more because I was more energetic. I improved my physical health, mental capacity, and self-organization.

Now you might think that the road to perfection requires the sort of discipline and regimentation that makes a person rigid and lacking in creativity—that the overachievers of the world become so structured that they lose all sense of spontaneity and feel they're doing you a favor by fitting you into their busy schedules. But that's not the case if you do it right. I discovered that what really makes people rigid, tense, and negative is the persistent problems they deal with—the limitations and frustrations that they struggle hopelessly to overcome. When they equip themselves with the knowledge and techniques to achieve mental freedom and allow themselves some creative space, the more unique, special, playful, and socially conscious they become. When you break loose from the restraints of poor time and personal management, the world very soon looks better.

My years of testing, developing, and improving are all summarized here for your benefit. By working through this book, you're laying the groundwork for a richer, more productive, and happier life.

What Were You Taught?

Before we take a closer look at improving brain functions such as logic, numeracy, analysis, and imagination, I want you to think about what you were taught about how your brain functions. Answer the following questions by marking the Yes or No checkbox for each. The goal is to create a permanent record of your answers so that you can look back at them later to see how much progress you've made after completing this program.

As part of your education—grade school, high school, or college—were you taught . . .	Yes	No
1. about the differences between the left and right brain?		
2. about the mathematical, memory, and learning potential of your brain?		
3. how your memory changes during learning?		
4. how your memory changes after learning?		
5. how to develop a surefire recall system?		
6. how to develop your listening ability?		
7. about the role of eye movements in speeding the learning process?		
8. how to dramatically improve comprehension?		
9. about shorthand techniques?		
10. how words and images interact in your mind?		
11. note-taking techniques that help you form mental pictures of your thoughts?		
12. how to prepare for and take tests?		
13. how to make reports, speeches, and presentations?		
14. methods for quickening the pace of mathematical calculations?		
15. techniques for analyzing arguments to detect flaws in logical reasoning?		

Most people answer yes to just a handful of these questions. Many people say no to all of them. Think of all the wasted potential in our early years. Think of the you that will emerge when you can answer yes to virtually every one of those fifteen questions after completing this program.

CHAPTER 1

Getting Up to Speed on Brain Basics

Everything we do, every thought we've ever had, is produced by the human brain. But exactly how it operates remains one of the biggest unsolved mysteries, and it seems the more we probe its secrets, the more surprises we find.

—Neil deGrasse Tyson

The brain is a complex organ encased in the confines of your skull. It processes sensory information and other input as well as controlling thought, memory, emotion, touch, physical movement, vision, heartbeat, blood pressure, breathing, temperature, digestion, and every other physiological process. Even before you're introduced to the formal study of mathematics, your brain performs thousands and perhaps even millions of calculations daily and instantaneously just to control your physical movement in the world.

While we rarely give it much thought, our brains are biological supercomputers consisting of about 100 billion neurons. Our capacity to store and process information and think is a function not only of those billions of neurons but also of the connections between them—the neural networks that form as we take in information and use and challenge our brains. These communication and data storage networks all rely on both electrical and chemical messaging systems that communicate faster than the speed of light—at the speed of thought.

In this brief chapter, I introduce you to brain basics, explain the differences between the left and right sides of the brain, and explain how to put the rest-activity cycle to work for you to make learning more efficient and make it feel less burdensome. Finally, I close the chapter by introducing you to the concept of the upper and lower brains.

The Tale of Two Brains

In recent decades, we have discovered that we actually have two brains—a left brain and a right brain, as shown in Figure 1-1. In California laboratories in the late 1960s and early 1970s, research was begun that was to change the history of our appreciation of the human brain, and which was to win a Nobel Prize for Roger Sperry of the California Institute of Tech-

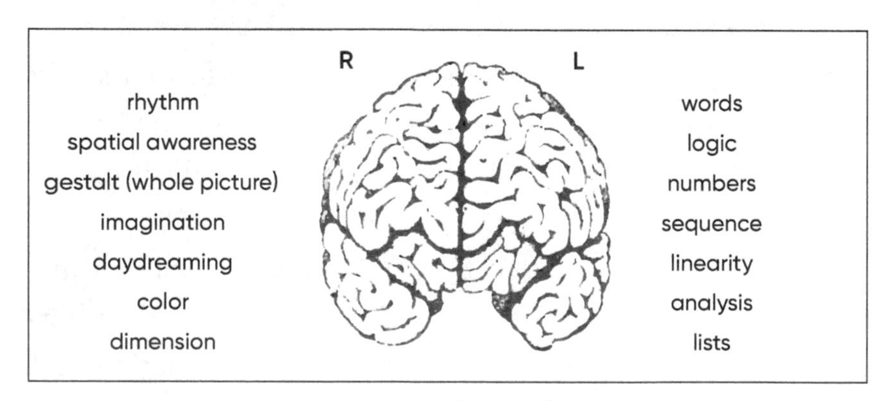

Figure 1-1: Front view of the two sides of the brain and their functions.

nology and worldwide fame for Robert Ornstein for his work on the brain waves and specialization of function, carried on through the 1980s by Professor Eran Zaidel of the University of California at Los Angeles.

In summary, Sperry and Ornstein discovered that the two sides of your brain, or your two cortices, which are linked by a fantastically complex network of nerve fibers called the *corpus callosum*, deal dominantly with different types of mental activity. In most people, the left cortex deals with logic, words, reasoning, number, linearity, analysis, and so on—the so-called academic activities. While the left cortex is engaged in these activities, the right cortex is in the "alpha wave" or monitoring state. The right cortex deals with rhythm, images, imagination, color, parallel processing, daydreaming, face recognition, and pattern or map recognition.

Subsequent research has shown that when people were encouraged to develop a mental area that they had previously considered weak, this development, rather than detracting from other areas, seemed to produce a synergetic effect in which all areas of mental performance improved.

Developing Both Sides of Your Brain

Despite the great body of evidence that has been presented over the last several decades, many people aren't sure that they really do have left-brain and right-brain capabilities. But they do; we all do. For example, let me ask you a question: can you speak and understand the language in which this book is written? Of course you can. And if you can, you have automatically used your left-brain skills of numeracy, logic, analysis, and sequencing. You've also used your right-brain skills of rhythm, imagination, and spatial awareness.

The topic of spatial awareness brings us back to the idea of where you find yourself physically when you have those spasms of imagination, those amazing bursts of ideas. Are you, like most people, driving, strolling in nature, or skipping stones on a pond? Perhaps you're in bed or the shower, or staring into a mirror. It's when you are relaxed and unhurried, restful, and usually alone that you have those flights of fancy, those amazingly imaginative bursts of ideas.

That takes us back to Albert Einstein. Einstein credited his imagination, his imagination games, and his right-brain skills with providing many of his greatest scientific insights, although we conventionally attribute these to the left brain. On one balmy summer day, when he was lying on a hill daydreaming, Einstein imagined himself riding sunbeams to the farthest points of the universe. Finding that no matter how far he went, he illogically returned to the surface of the sun, where he had started, he suddenly realized that the universe must be curved and therefore finite. And he realized that his previous logical training didn't go far enough. Inspired by this epiphany, he built the numbers, the equation, the words, and the imagery to describe it in scientific and mathematical terms, ultimately delivering to the world his new theory of relativity—a product of the synthesis of right-brain and left-brain activity.

When you think of a great artist, do you think of that person as being right-brain or left-brain dominant? Most people would answer, "right-brain dominant." Yet, many artists, in fact, virtually all of them, people we think of as wild and unstructured, make meticulous notes about tones and combinations of colors and the sequencing of colors that produce visual balance. When pop artist Andy Warhol died, his personal diary showed records of expenses for everything, including every cab ride, exactly to the penny.

So, when you say that you're a terrific musician but poor in math, you're simply describing one area you have chosen to develop and another to which you have given little attention. Yet you can always pivot to focus your mental energies on a discipline that more fully engages the opposite half of your brain. Even if you are left-brain or right-brain dominant, the less engaged half of your brain doesn't simply wither and die. Research shows that when people are encouraged to develop a mental area they had previously considered weak, half of the brain responds to the challenge. Rather than detracting from other areas of mental performance, the challenge produces a synergistic effect in which all areas improve. By reading this book and performing the exercises given here, you'll not only develop the areas in which you're strong, you'll dramatically enhance those areas in which you considered yourself weak.

Mastering the Rest-Activity Cycle

Before diving headfirst into this mental fitness program, first understand and appreciate the importance of achieving a balance between rest and activity. Our culture prides itself on phrases such as "nose to the grindstone," "busy as a bee," and "work your fingers to the bone." We tend to look at people who talk about quality of life as being somewhat lazy or undisciplined. Many of us disdain time off, but the brain can take only so much activity (anywhere from 20 to 60 minutes at a time) before becoming drained of oxygen and physiological resources. The brain, like the body, requires rest periods.

If you have ever had to sit through a long class, presentation, or meeting, or even a three-hour movie, you know that your brain begins to shut down well before the conclusion, and you can't just call time-out. Whoever or whatever is in charge continues to push forward, much as a jockey whips his horse on a final drive to the finish line. That doesn't work so well with the brain. To reap the maximum reward from a period of challenging brain activity, you need to master the rest-activity cycle.

When you allow the brain to rest, you give it time to recuperate, reorganize itself, and prepare for the next cycle of mental activity. Resting the brain does not necessarily mean doing nothing. You can take a brain break simply by switching from a left-brain to a right-brain activity, or vice versa. For example, if you've been working on math problems for a period of time (left brain), switch to a right-brain activity, such as daydreaming. Just look at what Einstein accomplished when he went off on his flights of fancy.

You can also choose to disengage *both* halves of the brain. We often do this in periods of sleep or meditation. Of course the brain continues to function. It continues to coordinate our heartbeat, breathing, and digestion. It processes all the day's inputs. It even solves problems. During these downtimes, the brain also repairs itself and detoxes. Just as your car produces waste products as it burns fuel, your brain produces waste products that must be removed to keep it healthy and support optimal function.

Physical rest is not the purpose for taking mental breaks during the day. The whole idea behind alternating mental activity with rest periods is to enable you to focus more effectively on whatever requires your concentration in the moment. In some ways, losing concentration is the brain's way of telling us it needs a rest. It's the brain's way of protecting itself from too much wear and tear and, in many cases, too much information all at once.

What happens if we don't give in to these lapses of concentration? What if we push ourselves beyond 60-minute spurts? Certainly you can, in the short term, ignore your brain's requests for rest. You may not achieve peak efficiency, but you can carry on without planned breaks. In the long term, however, you can't. In extreme cases, without rest, your brain becomes increasingly susceptible to suffering a "nervous breakdown" or physiological collapse.

Recognizing the Upper and Lower Brain

The left and right brain are probably familiar to you, but are you aware that the brain is also divided into an upper and lower brain? The upper brain is often called the *conscious brain* or *cerebral cortex*. It's the part of the brain we refer to when we tell someone to "put on their thinking cap." The upper brain deals with your intellectual activities. So when we talk about the left and right brain, we're really talking about the left and right *upper* brain. The lower brain, sometimes referred to as the *subconscious brain*, handles our everyday functions, such as regulating blood pressure, body temperature, digestion, and other "automatic" operations.

What is exceptionally fascinating is that the upper brain can control the lower brain. As a result, the mind can control the body. There are thousands of documented studies of people who have recovered from illnesses simply through the power of their minds. Likewise, people can literally worry themselves sick. For example, students who suffer from severe test anxiety can experience stomach upset, nosebleeds, or hives at the mere thought of a coming exam. According to some studies, approximately 80 percent of the back pain severe enough for people to consult their doctors can be attributed to conscious or subconscious emotional or psychological stress: their doctors can find nothing structurally wrong with their backs. These are all instances of the upper brain controlling the lower brain, either consciously or subconsciously.

In my Olympic coaching, we find that many of our most successful athletes practice *imaging*. In their mind's eye, they imagine every step of their performance and how they will win an event. Their upper brain uses its astounding power to control the enormous energy of the lower brain to produce their desired outcome. The slightest conscious doubt that they will succeed is often sufficient to completely undermine their performance. Legendary gymnast Simone Biles once said, "Mentally, I have to get my body and mind in the right place before I start the routine, but once into the zone, it's like I turn on a switch."

In the next chapter, you will have the chance to expand your introduction to the left- and right-brain functions. Once you do, you'll see how this program will make a dramatic impact on your performance in every area of life.

Don't underestimate the power of your subconscious mind

According to many experts, philosophers, and spiritual thinkers, the subconscious mind does far more than merely regulate and coordinate autonomous bodily functions. It is the part of the mind that connects us all to one another and to the omnipotent and omniscient force that permeates the universe. By transferring our conscious desires to our subconscious minds, we can harness the creative power of the universe and tap into the collective mind power of all living beings to bring to fruition whatever we imagine and fully believe is possible. Through the power of your subconscious mind, you can heal yourself and others, achieve incredible success in your career or business, live a richer and more fulfilling life, and even develop psychic powers such as clairvoyance and mental telepathy.

Several books have documented the power of the subconscious mind, including Joseph Murphy's best seller *The Power of Your Subconscious Mind*, Napoleon Hill's *Think and Grow Rich*, and Rhonda Byrne's *The Secret*. The authors present various techniques for transferring conscious desires to the subconscious, including repeating daily affirmations.

Even if you don't believe that the subconscious brain has metaphysical powers, you would be wise to nurture a positive mindset. Success in any area, including optimizing your brain power, requires a certain degree of confidence and certainty that you will be successful. The power of positive thinking delivers the confidence and certainty required. People don't generally succeed at anything when they're convinced they will fail. Often they don't even try.

CHAPTER 2

Taking Your Math Skills to the Next Level

Math is like going to the gym for your brain.
It sharpens your mind.
—DANICA MCKELLAR

You may be highly skilled at math or not. At one time, many schools discouraged girls from studying mathematics. Whether it was based on superstition or age-old assumptions about the female's place in society, there has been discrimination. Some of us may have been turned off to studying math because of how it was taught or because at a certain level, for example higher algebra or calculus, we no longer recognized its practical application in our lives. But being uninterested in math or discouraged from studying it does not mean that we lack a capacity for developing skills in this area.

In the past, it was assumed that certain people have mathematical ability and others did not, no matter how much help they were given. Now we know that if we don't do well in a skill such as mathematics, it's simply because we never allowed ourselves to develop in that area. Not because we can't: we can. In fact, I want to restate that you are capable of doing any form of mathematical calculation whatsoever.

For example, every day you perform some of the most astoundingly complex mathematical calculations and computations imaginable. Your eyes receive billions of tiny bits of information every nanosecond, from which you automatically calculate distance and speed. From that data, you assess the speed of an oncoming automobile and calculate with staggering speed and complexity whether or not you have time to cross the street without being hit. And you do it far more brilliantly than any known computer.

You are good at mathematics even if you don't know it. If you follow my techniques, you will improve your ability to add, subtract, multiply, and divide. You may not go on to become the brilliant mathematician you're capable of becoming, but you will begin to realize that you do have an innate aptitude for it. Achieving your potential is simply a matter of application.

Your Amazing Latent Mathematical Capacity

We now know that there are four main factors which determine mathematical ability (none of which is race or sex):

1. Techniques
2. Practice
3. Memory
4. The brain's basic ability

1. TECHNIQUES

Over the years, mathematicians have developed increasingly easy techniques for dealing with different forms of mathematical calculation. These are now more readily available, and those who use them are inevitably better "calculators" than those who do not. This chapter outlines, with examples, some of the more basic techniques.

2. PRACTICE

All good calculators, especially those considered great, have admitted that their skill was achieved not only by techniques but also by continual practice: none of them was born crunching numbers. As with other mental areas, the ability to calculate is a skill and requires practice to familiarize the brain with the various aspects of the task.

As you practice, your brain begins to develop neural networks, increasing your biological and physiological ability to perform math. The more you practice, the more effective and efficient these neural networks become.

3. MEMORY

Without exception, the great calculators have committed to memory many of the fundamental techniques and formulae necessary to their art. Previously, this was seen as an insurmountable barrier by those who thought that the memory had a limited capacity. Yet each one of us has an almost limitless memory, and if you use it properly, it will find all memory tasks easier, including mathematics. (For more about developing your memory, turn to chapter 4.)

4. The Brain's Basic Ability

Just as we underestimated the capacity of memory in the past, so we underestimated the general ability of the brain. It was assumed that some people were basically capable in mathematics and that others were not, no matter how much assistance they were given. Now, of course, we know that the brain has an almost limitless capacity, which ranges over all subjects, including the sciences and arts. Professor Ornstein's research on the left and right halves of the brain has also shown that each of us has a "mathematical" brain and an "imaginative" brain, and that our potential in each is basically equal. Any "disability" that we may have is probably due to neglect rather than to any innate lack of ability.

Addition

In the course of our math instruction, most of us have been taught to add columns of numbers right to left from the ones column to the tens column to the hundreds column and so on, carrying over amounts from one column to the next. For example, add the following column of numbers:

57
58
33
91
72
46
19
64

World record for mental math

Sanaa Hiremath holds the Guinness World Record for performing mental math—solving math problems in her head, without the help of a calculator, pen, or paper. She's 11 and was diagnosed with autism when she was two years old.

Sanaa failed second grade math. When her teachers tested her on math and told her to write the numbers from 1 to 20 on a piece of paper, she was unable to perform the task. She lacks the fine motor skills required to hold and use a pencil.

However, when her parents first introduced to multiplication problems, she was able to answer them instantly. As her parents and teachers presented her math problems of increasing complexity, she was able to answer those quickly and easily as well.

To set a new world record for performing mental math, she had to solve a 12-digit multiplication problem in her head in less than 10 minutes. She's now tackling math problems typically given to engineering students at MIT.

You probably started at the top right column and worked down: 7 plus 8 is 15; 15 plus 3 is 18, 18 plus 1 is 19; 19 plus 2 is 21; 21 plus 6 is 27; 27 plus 9 is 36, which, plus 4, is 40. Now you put down the zero, carry the 4 to the top of the second column, and continue in the same way. It's time-consuming, and you're still not done.

Among the many techniques for addition, the following four make the basic process considerably easier:

1. 10 packets
2. Complete 10s
3. Multiples
4. Splitting the numbers

EXERCISE 1: 10 PACKETS

Whenever adding long columns of numbers, always look for the 10-packets. For example, when adding

57
58
33
91
72
46
19
64

It's a waste of time to add them up by mumbling to yourself, 9 plus 4 is 13, plus 6 is 19, plus 2 is 21, plus 1 is 22, and so on.

It is far easier to link the numbers that make 10, giving you a series of 10-packets. By quickly and lightly striking through the 10 packets with a pencil, you do not lose track of the ones you have already packeted. Thus, in the right column above, the 7 and the 3 make a 10; the 8 and the 2 make a 10; the 1 and the 9 make a 10; and the 6 and the 4 make a 10, giving us an easy calculation of 40. In the next column, the 5 and 5 make a 10; the 3 and the 7 make a 10; the 9 and the 1 make a 10; and the 6 and the 4 make a 10, which, when we add the 4 from the other column gives us 440.

EXERCISE 2: 10 PACKETS CONTINUED

The 10-packet technique works equally well even if the numbers don't add to 10. Look at the next long column of numbers. Do the 10 packets. Use a pencil to cross the numbers out as you say them in order to prevent confusion. Try it:

93
28
32
86
61
17
44
22

In the ones column, the 3 and the 7 make a 10, the 8 and the 2 make a 10, and the 6 and the 4 make a 10, giving three 10s with 3 left over: 33. In the next column, the 9 and the 1 make a 10, the 8 and the 2 make a 10, and the 6 and the 4 make a 10 with 5 left over, to which we add the 3 from the previous column, giving a total of 383.

As you can see, you can easily double the speed of your addition simply by using the 10-packets technique. And, as you are about to see, addition techniques can make life easier and faster as well.

EXERCISE 3: COMPLETE 10s

You take the 10-packet technique one step further by combining two-digit numbers that form multiples of 10. This complete-10s technique can be seen clearly in the following example:

34
26
97
15
13
55

By blending these in an easy manner, we can break this column down into: 34 and 26 make 60; 97 and 13 make 110; 15 and 55 make 70. Then 60 plus 110 plus 70 equals 240.

As with 10 packets, complete 10s can often reduce by one half the time you spend on a lengthy calculation.

EXERCISE 4: MULTIPLES

Many people find it easier to combine similar numbers and multiply them rather than being confronted with a long list of different numbers. When you are faced with a long column of numbers, check down the entire column, seeing how many numbers are repeated. You can make your addition problem simpler by applying the multiplication tables to what initially appeared to be a large and threatening addition problem. For example:

7
8
6
1
5
7
7
9
6
5
1
5
9
9
8
7
8

Lightly marking off the repeating numbers, we find three 9s, three 8s, four 7s, two 6s, three 5s, and two 1s. The problem can therefore be written out in this form:

$3 \times 9 = 27$

$3 \times 8 = 24$

$4 \times 7 = 28$

$2 \times 6 = 12$

$3 \times 5 = 15$

$2 \times 1 = 2$

The addition then becomes a far simpler one, and the correct answer of 108 is more easily found.

The example given here is a fairly short addition problem, but the more numbers you have to add, the more useful this technique becomes.

EXERCISE 5: SPLITTING THE NUMBERS

The final addition method is called *splitting the numbers.* It can increase your speed of calculation of difficult additions by as much as five times and sometimes infinitely, because without it many people just give up.

With this method, you simply split the numbers into smaller chunks. Suppose you were given the following addition problems:

$$\begin{array}{r} 31 \\ +54 \\ \hline \end{array} \qquad \begin{array}{r} 425 \\ +379 \\ \hline \end{array}$$

You could solve them by mental arithmetic if you "split the numbers." All that this implies is that you separate the numbers into two smaller parts, making a "hard" addition comparatively easy. Once you have done this the addition can be done with amazing simplicity.

In the first example we split the numbers 31 and 54:

3	1
+5	+4
8	5

Seen like this, the answer 85 is more immediately obvious.

This splitting-the-number technique becomes even more useful with larger numbers. In the second example, again we split them into two more easily manageable numbers:

42	5
+37	+9
79	14
+1	-10
80	4

The first simple addition gives us 79. The second gives us 14, from which we carry the 1 to the first addition as a single group of 10. Putting the numbers together gives us the answer: 804.

This technique and the others presented in this chapter can all be improved by practice.

Subtraction

There are three main techniques for making subtraction easier:

1. Addition
2. Splitting the numbers
3. My quick subtraction technique

ADDITION

Imagine solving a subtraction problem with addition. Take an example like the following:

```
 7596
-4779
-----
```

Most people would go through a process something like this: "9 from 6 won't go, so borrow 1 from 9 to make 16; 9 from 16 is 7; 7 from 8 is 1; 7 from 5 won't go, so borrow 1 from 7 to make 15; 7 from 15 is 8; 4 from 6 is 2."

Rather than performing a fairly laborious process, it is far easier to apply addition to the subtraction problem. Before you start, "prepare" the numbers. If a number above is smaller than the number below, you add 10 to it, and add 1 to the lower figure in the next column to the left. The subtraction is then done by reciting the number you need to add to each lower number in order to make up the number above it.

Thus, in the example given, the subtraction will be done in the following way: the 5 and the 6, being smaller than the 7 and the 9, will be made into 15 and 16:

```
 7    15    9    16
-5    -7   -8    -9
--------------------
 2     8    1     7
```

By using this method, you will have arrived at the correct answer of 2,817 in a more rapid manner than by the traditional method.

In long subtractions, you can save an enormous amount of time with this approach of preparing the numbers before you start to subtract and then "adding" to find the correct answer.

SPLITTING THE NUMBERS

As with addition, subtraction can also be done more easily if larger numbers are split up into smaller ones.

Imagine that you are given the following subtraction problems to solve:

97	154	528
-32	-42	-212

In the first example we split the number 97 and 32 as follows:

9	7
-3	-2
6	5

Seen like this, the answer, 65, is once again more immediately obvious.

The two more "difficult" examples can be similarly split up, so that the answers are almost immediately available.

15	4
-4	-2
11	2

52	8
-21	-2
31	6

In subtractions where the lower number has individual numbers in it that are larger than the individual numbers they are being subtracted from, simply make the appropriate adjustments of adding 10s and giving 1s to the next column when splitting the numbers. To subtract 247 from 393, prepare the numbers like this:

39	13
-25	-7
14	6

MY QUICK SUBTRACTION TECHNIQUE

In a traditional math setting, you are instructed to subtract starting from right to left. For example, to subtract 67 from 100, you first subtract 7 from 0, which of course you can't do. So you try to borrow from the next number over, which is also 0. Next, you move over to the 1 and give it to the 0 on its right to make it 10. Then you borrow 1 from that 10 to make the 0 on the right 10. Now you can subtract the 7 from the 10, leaving 3. Since you borrowed 1 from the 10 that was over the 6, you're left with 9. 6 from 9 is 3. So the answer is 33.

Let's take the same problem, 67 from 100, and subtract from the left. First, change the first 1 and 0 in 100 to 9 and change the second 0 to 10. Now subtract left to right: 6 from 9 is 3, and 7 from 10 is 3, giving the answer 33. Pretty fast, huh?

9	10
-6	-7
3	3

This technique works with numbers like 100, 1000, 10,000, 100,000, or 1 million. You change the 1 and 0 to 9 and every other 0 to 9, except the last 0, which becomes a 10. This makes it very easy for us, because we normally read from left to right. Here's another example: Subtract 624 from 1,000.

Change the first 1 and 0 in 1,000 to 9, the second 0 to 9, and the last 0 to 10, then subtract from left to right:

9	9	10
-6	-2	-4
3	7	6

The answer? 376.

Subtract 792 from 10,000:

9	9	9	10
-0	-7	-9	-2
9	2	0	8

As you can see, when the number you're subtracting from has more digits than the number you're subtracting, you position the bottom number below the top number so that their rightmost digits are aligned, then you add zeros (one in this case) to the left of the bottom number before doing your subtraction. In this case, we subtract 0792 from 10,000.

Another way of looking at this method is to say that when subtracting from, say, 1,000, we change that number to 999 for the moment, and then add 1 to compensate.

9	9	9	9
-0	-7	-9	-2
9	2	0	7
			+1
9	2	0	8

Here's one more: Subtract 7,438 from 10,000:

9	9	9	10
-7	-4	-3	-8
2	5	6	2

This is a great way to make quick calculations. With just a little practice, you can amaze your friends. You'll eventually get to the point at which you can give an answer almost as soon as they say the number to be subtracted. This is also the beginning of proving to yourself that you can do arithmetic not only well, but quickly.

Try this technique both by writing your numbers and answers down and then by doing them in your head.

Multiplication

In multiplication, there are two especially useful techniques:

1. multiplying by 5
2. multiplying by 11

MULTIPLYING A NUMBER BY 5

To multiply by 5, we multiply by 10, then divide by 2. So to multiply 5 times 84,580, we simply add a zero (which is the same as multiplying by 10) to get 845,800; then we divide by 2 to get the answer, 422,900.

This method saves a lot of time and is even more useful when combined with the "group vision" technique for dividing numbers, as explained in the division section.

MULTIPLYING A NUMBER BY 11

To multiply a two-digit number by 11, add the two digits together and put their sum in the middle.

For example, if we want to multiply 72 by 11, we split the 7 and 2 and put in the middle the sum of 7 and 2: 9, giving us 792 as the answer.

If the sum of the digits comes to 10 or more, simply add 1 to the left digit. For example, if we wanted to multiply 85 by 11, we would split the 8 and the 5, add them to give 13, giving us the answer 935.

In the same way that this technique suddenly becomes easy, there is an enormous range of quick- calculation techniques for multiplying longer numbers by 11, multiplying numbers by 15, 25, 50, 75, 125, and others. For those who are exceptionally interested, there are also techniques for the cross- multiplication of large numbers.

From the simple techniques demonstrated so far, it should be increasingly clear that mathematical ability depends on knowing how to do it—and there are many clever methods.

Division

Two easy methods immediately present themselves for the following types of division problems:

1. Division by 2
2. Division by 5

DIVISION BY 2

To divide numbers by 2, apply "group vision." This means splitting the number into easily manageable sections, as with addition and subtraction. For example, to divide the number 6,728,544 by 2, we simply split it up in the following way:

6 72 8 54 4

Each of these numbers is easily divided by 2, and by running through them in order we get the answer 3, 36, 4, 27, 2 and string them together to get 3,364,272.

This brief example reveals that anyone who wishes to become a rapid calculator must learn to scan numbers before getting involved too deeply in them. When you are dividing by 2, split up the number by means of small strokes or dashes (as you progress, you will be able to split up numbers quickly at a glance, without the strokes or dashes). This group vision, while applying in particular to division by 2, is also a useful general principle in quick calculation.

DIVISION BY 5

To divide any number by 5, divide by 10 and double it.

For example, 823 divided by 10 equals 82.3 multiplied by 2, equals 164.6.

As with the other mathematical procedures, there are many different techniques for dividing. The ones shown here are simply indications of the easier paths that are available.

By learning these new number skills for addition, subtraction, multiplication, and division, first you will have shown a willingness to open your mind to new methods, and second, you will have come to the realization of just how effective certain techniques are: they work. Third, you will now be able to use your new skills in many real-life situations, where quick and accurate calculations are helpful and necessary.

Finally, you're probably now beginning to think, "Hmm, maybe I can stretch myself a bit. Maybe I am a mathematical genius." A moment ago, you may have been reluctant to say the numbers aloud or even mumble them to yourself during a calculation. Now you may be doing it without a second thought. The point is that if you can just imagine the numbers and see them more powerfully in your mind's inner eye, you will be able to calculate faster. The individuals who are expert calculators use this imagination system because it involves the right-brain skill of imaging. Use it, and your mathematics will improve dramatically.

In other words, it's a distinct advantage in the speed of mathematical calculation to be able to incorporate both sides of your brain. And we are now finding that it is easy to establish new learning habits. In fact, we now know that an older brain, if it is trained well, can develop new skills faster than any child's brain.

CHAPTER 3

Thinking Rationally and Spotting Logical Fallacies

Education has failed in a serious way to convey the most important lesson science can teach: skepticism.

Most people consider themselves rational, but logic often has very little to do with their decision making. For example, people frequently accept as fact what they hear, see, or read in the media, simply because they trust that media channels have carefully vetted the news for accuracy. People often fail to realize that everyone, including you and me, is biased. Each of us has deeply ingrained beliefs and a tendency to believe what we hope to be true. Media channels and reporters are not immune to bias.

The ability to apply logic can be of tremendous value in making decisions in your personal or professional life. For example, suppose you're trying to decide whether to get vaccinated to protect yourself from infection with a potentially lethal virus and spreading it to others. You've already had the infection and recovered from it. Health authorities are advising that everyone get vaccinated, but you've seen studies showing that natural immunity from having the infection is superior to the immunity granted by vaccination. Other studies have found that vaccination risks are higher for those who have already had the virus and recovered from it. What do you do? Do you follow the advice of the government's medical authorities, or do you draw your own conclusions from the available evidence?

Whether you realize or not, the world is full of misinformation, misleading information, and faulty logic. Unfortunately, problems arise when authorities try to stem the tide of misinformation. Often human bias influences the criteria for differentiating between what's true and what's not. The people in power end up deciding that whatever they believe (or want you to believe) is true and that whatever they don't believe (or want you not to believe) is false.

The only solution that seems to work effectively is a combination of free speech and education. We need to allow for the free flow of information along with open debate and teach *everyone* how to distinguish between fact and fallacy for themselves. And we need to approach every claim and every statement with a healthy dose of skepticism. In this chapter, we will point you in the right direction and provide several examples and exercises for reinforcing logical thought processes.

Analyzing Logic

A logical argument is one in which, if the basic facts or premises are true, the conclusions that follow must be true.

Checking the facts on which an argument is based is usually easy enough: you trace them back to the source or conduct your own investigation. You may even question the source to determine how the facts were established (although sources are not always reliable).

After verifying the facts, you need to ask whether the argument based on those facts is legitimate: does it lead to a logical conclusion? In many cases, an argument that appears logical on its surface can arrive at a totally absurd conclusion. Here's an example of a false argument that could be used to support a prejudiced belief:

Premise: All foxes have tails.

Statement of (questionable) fact: The animal I saw running into the forest has a tail.

Conclusion: The animal I saw running into the forest is a fox.

Many people have been led astray by arguments that appear to be logical but are not.

There are two major forms of logical presentation and misrepresentation, and they can be stated as follows:

1. All As are B
   All Bs are C
   Therefore, all As are C

2. All Bs are C
   A is a C
   Therefore, A is a B

One of these presentations is correct; the other is not. One way to apply logic is to do it visually, using circles to depict the groups A, B, and C.

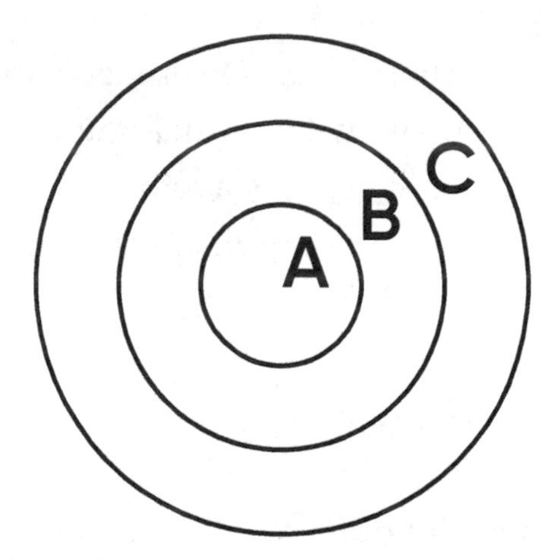

Figure 3-1: All As are Bs, All Bs are Cs, so all As are Cs.

A simple diagram of the first presentation shows that if the premise or initial statement is correct, then the final statement will also be correct (see Figure 3-1). If we take the group B, and make it a circle, then the group of As will be a smaller circle within the group of B, because all As are B. If all Bs are C, then C will be an even larger circle than B, and *all* As will therefore be C.

The soundness of this reasoning can be illustrated by a simple example:

All ants (A) are insects. (B)
All insects (B) are six-legged. (C)
Therefore, all ants (A) are six-legged. (C)

Thus this form of argument is correct if the premise is true. If the premise is *not* true, this form of argument not only collapses but often leads to an absurd conclusion, as can be seen from the following example:

All berries are good to eat.
The deadly nightshade is a berry.
Therefore the deadly nightshade is good to eat.

Although this argument is in correct *logical* form, it is untrue because its first premise is false. So when listening to, or reading, arguments, it is necessary to make sure that the premises are accurate.

The next thing to check is whether the statement follows a logical train of thought. The second logical presentation mentioned above is an example of erroneous reasoning, as illustrated in Figure 3-2. Here again, C represents the large group, and because all Bs are C, B can be shown by a smaller circle inside. A is also a C, but A could be either inside the B circle or, significantly, outside it. To reach the final stage of the argument by saying, "Therefore A is a B" is to state something that may be true but is not necessarily true. It is therefore a false argument.

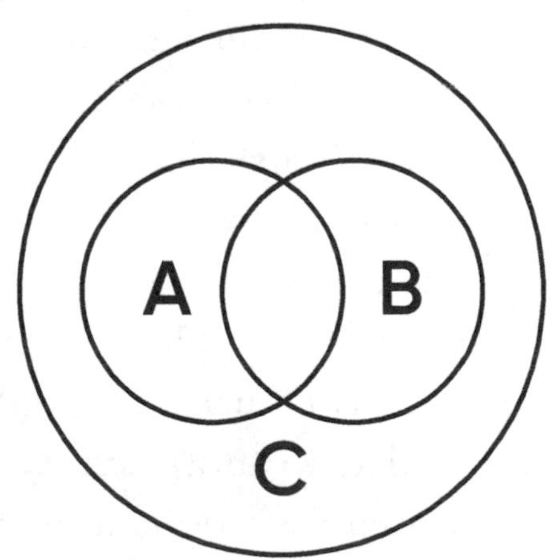

Figure 3-2: All As are Cs, All Bs are Cs, but not all As are Bs.

The incorrectness of this form can be seen through the following example: All singers (A) are dancers. (C)

All actors (B) are dancers. (C)

Therefore, all singers are actors.

This type of argument is common in discussions of politics, race, and religion. Knowing logical structure, we can help prevent such discussions from becoming unnecessarily argumentative and unproductive.

Defining Terms to Alter Their Meaning

Many communications and discussions become bogged down because the meanings of the key words in the argument change, sometimes subtly and almost imperceptibly, as the argument progresses. This is especially true of discussions revolving around concepts such as "peace" and "good and evil," as well as in discussions about race, religion, politics, and philosophy.

In such discussions, one should of course attempt to define the words. At the same time, we must realize something about the basic nature of words: that rather than having an absolute definition, each word can have a great variety of additional meanings. Each person will have different associations for each word. It is therefore important in discussion to find out exactly what meanings the other people have associated with their key words. People are often surprised to find that within a group it is they and they alone who have given a word a certain meaning, which they have assumed to be common.

To bring this point home, ask a friend to think of the first six words that in his or her mind are associated with words like "run," "God," "happy," and "love," and then compare them with your own list of the first six words that pop into your mind. The disparity will be both surprising and illuminating. I've done this exercise hundreds of times, and so far no two people have had identical associations for *any* given word.

If every discussion started with participants agreeing to the definitions of key terminology, we would have far more understanding and far less conflict in the world. Agreeing even on subtle differences in the meanings of words and the feelings associated with those words can help prevent them from being interpreted as evasive or argumentative. As a result, those same words can be used constructively, to increase understanding rather than causing confusion and discord.

Arguments Using Biased Statistics

You have probably heard the claim that "numbers don't lie." While that is certainly true, people often use numbers to spin the truth (misrepresent the facts) in a way that is favorable to the point they are trying to make. For example, two newspapers—one favorable to the government in office, the other unfavorable—reported on unemployment. The first, in a major headline, stated: "Unemployment Total Stays Steady." The first paragraph of the story read as follows:

> Unemployment remained practically static during the past month, when the total number of unemployed rose by 3,972

> to 601,874 on November 9, representing an unchanged percentage of 2.6.

The other paper used the banner headline: "Unemployment May Reach 700,000." Beneath this heading was the lead paragraph:

> Little by little the unemployment figures are creeping up again. This month they have once more topped 600,000, which means they are the worst November total for 30 years.

The conflicting stories even used different graphs to drive their points home visually (see Figure 3-3). As you can see, the graph on the left shows very little change in unemployment, while the graph on the right apparently shows a dramatic difference in unemployment from October to November.

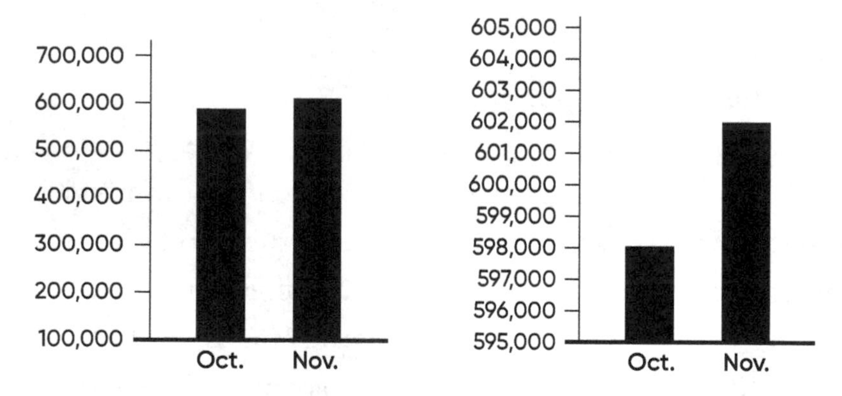

Figure 3-3: The same statistics can be used to tell vastly different stories.

Apart from a large number of logical dishonesties (can you spot them all?), these examples illustrate the selective use of statistics in order to bias the reader toward a certain viewpoint.

In the first article, the newspaper wished to give the impression of stability. All the possible figures were probably analyzed until one was found that showed very little change—the percentage.

The newspaper unfavorable to the government likewise found one specific statistic that made the issue look particularly serious.

This example helps to illustrate something more about statistics than the oft-quoted statement that "numbers don't lie." It shows that when you know what's behind the statistics presented, you will have a fairly accurate awareness of the situation and will also know what bias the presenter of the statistics has.

This approach to statistics makes them far more interesting: not only do they present numerical information, but they also add to the reader's knowledge of the motives behind the presentation of the material.

There are many ways in which figures can be purposely selected and biased, ranging from simple and false assumptions about averages to the more sophisticated manipulation of graphs to make good situations look bad and bad situations look better than they really are. If you are interested in getting to the "truth behind the numbers," either take a basic course in statistics or read up on statistics on your own. As soon as you have a fundamental understanding of statistics, hunting down logical fallacies can be an enjoyable and rewarding pastime.

Recognizing Other Types of Logical Fallacies

Over the course of history, people have developed many ways to bend the truth in favor of what they believe (or want others to believe). I've covered three so far. In the following list, I provide an overview of several other types of logical fallacies:

- **The correlation/causation fallacy:** When two events occur close to one another in time, people often assume that one event caused the other, which isn't necessarily true. For example, the United States elects a new president, and soon thereafter inflation drops. The new president argues that this is proof that his economic policies have been a resounding success, even when his policies weren't in effect at the time.

- **The bandwagon fallacy:** According to this fallacy, something is true when most people believe it to be so. Politicians who make their decisions based on popular polls are particularly susceptible to the bandwagon fallacy. As your parents probably advised when you were growing up, "Just because everyone else is doing it doesn't make it right." Likewise, just because most people believe something doesn't make it true.

- **The anecdotal evidence fallacy:** The anecdotal evidence fallacy involves basing a conclusion on a very limited data set. It's often used by people selling nutritional supplements to claim the effectiveness of their products when they have no large-scale, double-blind, placebo-controlled studies to prove it. For example, during the COVID-19 pandemic, people were desperate for natural cures and preventive supplements, and doctors often posted about the clinical success they had with treatments they tried on their patients. While clinical evidence does hold some weight, it doesn't equal the weight of a large, well-structured lab study.

- **The straw man fallacy:** This tactic involves misrepresenting an opponent's position and then attacking that position instead of what the opponent truly believes. If you follow the immigration debate, you can find plenty of examples of the straw man fallacy. On the right, politicians accuse those on the left of embracing open-border policies, while those on the left often accuse politicians on the right of being anti-immigrant. Accusing an opponent of taking a radical stance on an issue makes it much easier to argue against the radical position instead of the more nuanced view the person may have.

- **The false dilemma (or either/or) fallacy:** The false dilemma fallacy presents the audience with only two choices: the truth must be either A or B. Then it attacks one of the choices, leaving the other choice as the only possible alternative (through the process of elimination). However, the argument fails to acknowledge other possibilities. Sticking with the immigration debate, a false dilemma would be to argue that you must either support building a wall or support an open-border policy, when in fact someone can be against building a wall while wanting to see immigration laws enforced in other ways.

- **The appeal to authority fallacy:** This fallacy involves assuming that something is true merely because an expert or an authority on the topic claims that it is true. For example, if 7 out of 10 dentists recommend toothpaste A over toothpaste B, that doesn't necessarily mean that toothpaste A is better at preventing cavities. Empirical evidence, such as from a large-scale, double-blind, placebo-controlled study, would be required to prove that.

- **The slothful induction fallacy:** Slothful induction occurs when an argument ignores evidence that contradicts its conclusion. For example, the preponderance of scientific evidence supports the theory that global warming and climate change are caused by human activities, including the burning of fossil fuels, but some who refuse to accept that claim will focus on the sparse evidence that calls it into question.

- **False analogy:** A false analogy is a comparison between two things that are not similar. For example, some people like to argue against the theory of evolution by saying something along the lines of, "The probability of life originating from accident is comparable to the probability of the unabridged dictionary resulting from an explosion in a printing shop." Even though the theory of evolution is still open to debate, evolution, a biological process, isn't anything like a printing press.

Recognizing your own bias

In addition to being on the lookout for misinformation and faulty logic, we need to become aware of our own biases, which can distort our thinking, influence our beliefs, and sway our decisions from the inside. Here are some common forms of cognitive bias:

- **Confirmation bias** is the tendency to give more weight to evidence that supports what we believe to be true.
- **Hindsight bias** is the tendency to believe that we knew all along what the outcome of an event would be from the very beginning.
- **Anchoring bias** involves being influenced most by the first piece of information presented or our first impression than by evidence that follows it.
- **Self-serving bias** involves taking the credit for everything when it goes right and blaming others when it goes wrong.
- **Optimism bias** is the tendency to overestimate the probability of positive outcomes when we're in a positive mood.
- **Pessimism bias** is the tendency to overestimate the probability of negative outcomes when we're in a negative mood.
- **Sunk cost bias** is the tendency to believe that our investments will produce a positive return, which often leads to continuing to pour time, money, or effort into something that's a lost cause.
- **The Dunning-Kruger effect** is the inability to recognize one's own incompetence or lack of ability in a certain area.
- **False consensus** involves overestimating how much other people agree with our position or approve of our behavior.

Making Logical Judgments and Decisions

Before accepting the logic of any argument, examine the following factors:

- Your own emotions or bias in favor or against a position or the person presenting the argument.
- The presenter's credibility or whether the presenter is an authority on the topic.
- The credibility of the source that provided the evidence on which the argument is based.
- The train of thought (logic) being used to lead from evidence to conclusion.
- Whether the presenter has a bias or an agenda to promote or is emotionally invested in the topic. For example, are you reading a news report or an opinion column? One is designed to win you over to a point of view, whereas the other is supposed to be totally objective. Unfortunately, many times even the news that's presented as being entirely objective is incredibly biased, bordering on or even crossing the line into propaganda. I've seen prestigious scientific journals use emotional terms, such as "sophisticated" and "elegant," to describe experiments they supported and terms, such as "haphazard" or "sloppy," to describe work with which they disagree. It's essential to discriminate between that which is logical and that which is emotional and to examine the veracity of the evidence and its source.

Keeping emotion out of face-to-face interactions can be especially difficult. For example, if your job involves negotiation, you may find that the parties involved, including you, can get frustrated and angry or even resort to using anger as a negotiating tool—to bully the other parties into accepting terms they would otherwise reject. One way to respond to anger is with anger, fighting fire with fire, but when you sink to that level, you may not like the outcome, or sometime later, you may regret letting the other person pull you down to their level.

One of my students related an incident in which he was in an intense discussion with a client, who started yelling at him. "You're a boob!" the client screamed. "That's the dumbest thing I've ever heard! I don't know why I'm wasting my time talking to you!" My student responded by smiling apologetically, telling his client that he was sorry he gave that impression of himself, and asking for an opportunity to correct it.

Some people would say that my student backed down, but that's not how he saw it. "That's not backing down," my student, a savvy businessperson, explained. "That's disengaging, defusing the situation, so that my client could make a more rational business decision. If I started yelling at him, nothing would've been accomplished. Instead, he settled down, mumbled an apology for screaming, and we finished our business. You shouldn't let emotion get in the way of doing business. You should harness it and use the energy it gives you to your advantage."

While emotion (the opposite of rational thought) can undermine productive decision making, sometimes trying to be too logical can also stand in the way of progress. You have probably heard the expression "analysis to the point of paralysis." It refers to the frequent mistake of overanalyzing a situation to the point of inaction or indecision. You can waste hours and even days mulling over a decision and end up never making it.

This happens to students who struggle with exams, and it also happens to businesspeople who have been known to spend hours trying to figure out simply where to start. People make the mistake of thinking of their choices in terms of either-or when a third option is available—the option to do nothing, to back out, to waffle, to not decide.

That third option is not always viable. When you are facing a situation that requires action, you must do something. However, by realizing that you have that third option and that it is not viable, you can force yourself into making a choice. You may be looking at two choices that are so close that it really doesn't seem to matter which choice you make, so you just choose one and move forward with it.

Some people resolve the dilemma by flipping a coin, which may seem to be a random and haphazard approach, but it can actually be a very creative solution. Sometimes after reexamining the option that the coin flip has indicated, people realize they don't want to choose it, which helps them select the alternative. Developing your ability to take a sideways look at things enables you to find new methods of doing business, having relationships, and enjoying your life.

We are all far more creative than we give ourselves credit for.

Infuse Logical Thinking with Creativity

Have you ever fidgeted with a paper clip? Most people have, bending it into different shapes. A paper clip is designed for one sole use: to hold papers together. But people have devised all sorts of clever uses for paper clips. To engage your left brain more in the logic of problem-solving, which is typically a right-brain activity, devise your own clever uses for a paper clip:

- ____________________
- ____________________
- ____________________
- ____________________
- ____________________
- ____________________
- ____________________

Here are a few clever uses that others have suggested:

- Replacement pull tab for a zipper
- Temporary replacement pin for the hinge on a pair of glasses
- Key for unlocking interior doors
- Fishhook
- Bookmark
- Fingernail cleaner
- Tie clip

Not bad for something so silly, huh? This program is all about challenging and stretching yourself, using your brain in new ways you may have never imagined. If you allow yourself to grow, you're going to manage information—and yourself—with an exciting new competence and perspective. As we continue, we'll refer back to many of these concepts from time to time in order to strengthen and reinforce them. Just give yourself the chance to learn.

Activities for Continued Self-Improvement

To keep your mind analytically sharp, engage regularly in one or more of the following activities:

1. Create a scrapbook of particularly outstanding or amusing examples from articles and news stories illustrating the various forms of logical fallacy.
2. In personal and group conversation, "set" yourself to remain continually on the lookout for flaws in arguments. This does not necessarily mean stopping somebody at the end of every sentence to point out his errors; just remain more alert, listen more carefully, and examine more closely the reasoning behind the conclusions that people draw.
3. Occasionally (or continually) examine your own methods of presentation and communication. You will often find the most useful information about your own logical weaknesses during times of extreme anger (when you're being irrational).
4. If you have friends or acquaintances who are interested in improving their logical and analytical abilities, make a pact to check each other whenever errors occur. This will keep you alert both as presenter of information and as analyst of information being presented to you. Young children, once they are shown logical fallacies, are often extremely quick to pick them out. Delightful games can be played (with scoring if desired) with television and radio newscasts and newspapers. Members of the family score points for being the first to recognize logical fallacies in the information they are absorbing.

5. Buy newspapers and magazines that challenge your own viewpoints and compare them with the newspapers and magazines you normally buy. Nearly every publication has an editorial bias. By comparing two publications with opposing biases, you can identify each publication's bias more easily. You can perform this same exercise by tuning into media sources that have an opposing slant, such as CNN and FOX News.

By engaging in these activities, you will become more confident, relaxed, and communicative with words. You will be more able to deal with the constant barrage of spin from the politicians of all sides and from media outlets.

CHAPTER 4

The Care and Feeding of Your Brain

For all aspects of memory, keep yourself physically fit.
Healthy mind, healthy body, healthy body, healthy mind.
Your memory needs oxygen as fuel, so why not feed it often?

—TONY BUZAN

The brain is perhaps the most neglected organ of the human body. We educate it, train it to perform specific tasks, and entertain it, but most of us don't pay much attention to brain nutrition, and we frequently fail to give it the rest and relaxation it needs for optimal health and function. As a result, our brain performance suffers both in terms of memory and cognition: our brains don't work as effectively and efficiently as they can, and they fail to achieve their full potential.

Neglect your brain no more. In this chapter, I provide the guidance you need to take better care of your brain. Here you discover the four essential foods it needs to thrive, the value of dream hobbies, the nature of your memory, how to manage the information in your memory more effectively, and how to organize your life in chunks to prevent overload.

But before you dive brain first into this chapter, remind yourself of your incredible human potential. You can do whatever you put your mind to, and it doesn't even require much conscious effort. For example, as a young child, you performed one of the most astoundingly difficult tasks imaginable: you mastered a language. Mastering a language means that you have an inherent understanding of rhythm, mathematics, music, physics, linguistics, spatial relationships, memory, creativity, logical reasoning, and thinking—a total integration of right and left cortical skills. And you learned your first language almost effortlessly, simply by hearing it and through regular practice (trial and error) over a few years' time. If you can do that, you can do anything!

Brain Food

I often talk about the brain as if it is a separate entity apart from the body, but it is also a part *of* the body. It's not merely a fancy calculator. It's a biological/physiological organ that requires proper care and feeding. To support its optimal health and function, the brain needs the following five essential ingredients:

- Nutrition
- Oxygen
- Information
- Love and affection
- Sleep

NUTRITION

Your brain, like the rest of your body, relies on the nutrients in food for fuel and for proper growth and development. Poor nutrition can impair both health and function, leading not only to poor cognitive performance but also to mood irregularities and mental and behavioral disorders. To support brain health and function, consume a balanced diet consisting of the following foods and fluids:

- Vegetables, especially green, leafy vegetables such as spinach, kale, and collard greens, are rich in brain-healthy nutrients like vitamin K, lutein, folate, and beta carotene.
- Healthy fats include those from cold-water fish (salmon, tuna, mackerel, herring, trout, sardines), organic olive oil, avocados, nuts, and high-quality omega-3 supplements. The human brain is nearly 60 percent fat, and omega-3 fatty acids are major building blocks of the brain.
- Healthy proteins are used by the body to make enzymes, hormones, and other chemicals that affect brain function and build and repair tissue. The healthiest sources of proteins are organic meats, fish, eggs, soy, legumes, nuts, and seeds. The problem with most nonorganic protein sources are the hormones (especially in meats) and pesticides they typically contain.
- Berries contain flavonoids, which can help improve memory.
- Green tea seems to activate parts of the brain linked to working memory.
- Turmeric and its active compound, curcumin, have strong anti-inflammatory and antioxidant properties. Curcumin

has been shown to increase levels of brain-derived neurotrophic factor (BDNF), which is like a growth hormone for the brain. Turmeric has been linked to improvements in memory, focus, motivation, mood, and sleep, and it may help to alleviate depression and anxiety.

- Dark chocolate contains flavonoids and antioxidants, which support brain health, along with caffeine, which acts as a brain stimulant.
- Water is essential for every cell of your body and nearly every physiological function, including circulation and detoxification. Dehydration can impair cognitive performance and memory, depress one's mood, and cause structural changes in the brain that show up on functional magnetic resonance imaging (fMRI). Insufficient volumes of cerebrospinal fluid and blood over time may increase the risk of Alzheimer's disease and vascular dementia.

In addition to supporting your brain with healthy nutrition, avoid foods and substances that are potentially harmful, especially the following:

- Sugar and sweets, including sweet drinks and their diet equivalents, especially those containing aspartame. One study linked diet drinks to three times more strokes and triple the likelihood of developing dementia.
- Refined carbohydrates, such as those in white flour.
- Trans fats, such as those in commercially baked goods, snack foods, margarine, fried foods, and hydrogenated or partially hydrogenated vegetable oils.
- Smoking or vaping. Anything other than fresh, unpolluted air will subject the brain to toxins.

- Alcohol in excess of a few drinks a few days a week.
- Marijuana/cannabis.
- Fish high in mercury. Mercury tends to concentrate in fish that are higher in the food chain, so you're generally better off eating smaller fish, such as sardines, instead of larger fish, such as tuna and swordfish.

OXYGEN

Although a developed human brain is somewhere between 1 and 3 percent of your total body weight, it requires about 20 percent of your oxygen intake. A mere 3–6 minutes without oxygen is enough to cause brain damage, but any restriction of oxygen to the brain can impair its health and function.

Many things you take for granted are really very complex calculations. Billions of your brain cells constantly undergo staggeringly complex electrochemical changes to process the information you're always sorting and receiving. It's been estimated, for example, that the entire network of the world's telephone systems, if properly compared to your brain, would occupy a part of it that would be the equivalent of an ordinary garden pea.

Your brain needs oxygen for energy, and the best way to provide it is to be aerobically fit. To ensure that your brain is properly oxygenated, improve your breathing and blood circulation through exercise—a combination of aerobic exercise (biking, walking, jogging, rowing, skipping rope) and strength training (lifting weights or doing resistance exercises, such as pushups, sit-ups, and pull-ups). Everything your gym teacher told you was true. If you can get your heart pumping at somewhere around 120 plus beats a minute for

20–30 minutes at least three times a week, you will build and maintain the cardiovascular system your body needs for healthy oxygenation.

You can also do cross-country skiing, disco, or any other kind of strenuous dancing, intense hiking or climbing, or active and prolonged lovemaking. Loving and lovemaking involve all the senses, provide wonderful toning and exercise for the body, and give the brain one of its four important foods.

Anyone who makes a conscious effort to improve his or her aerobic conditioning will take a major step in overall self-development. You'll find increases in intelligence and creativity, reduction in stress, and increases in stamina. It's almost redundant to say it, but I will: a physical fitness program is critically important to your well-being and life expectancy.

Remember the Latin phrase *mens sana in corpore sano,* meaning *healthy mind, healthy body.*

INFORMATION

Just as your body loses tone without exercise, your brain loses tone without information, intellectual activity, and challenges. The fact that you are reading this book and working through the exercises shows that you recognize the importance of keeping your brain active. The more you use it, the better it becomes. Use it or lose it.

The old assumption was that the brain automatically declined with age, no matter what you did. It reached a peak, it was thought, somewhere between the ages of eighteen and twenty-four, and then began a steady, slow deterioration, and toward the end sometimes a fast one.

The ability to retain and recall information, perform numerical activities, develop vocabulary, and be creative were all thought to diminish with age. Professor Mark Rosenzweig and many others have laid that old horror theory to rest. It's not true. Rosenzweig and others, including me, have demonstrated conclusively that if your brain is stimulated, no matter what your age, it will continue to develop new neural networks. It will become more biologically complex and sophisticated the older it gets.

Plenty of nonscientific information also supports the claim that the brain actually improves with age. Michelangelo was still creating great art and writing when he was eighty. Haydn wrote much of his most beautiful music towards the end of his life. And Picasso was creating well into his nineties.

One mode of information acquisition that is very important for left-brain development is vocabulary building. It's been demonstrated that well-developed vocabularies correlate with business, economic, intellectual, social, and personal success.

Why is there such a high correlation? Think about it. A highly developed vocabulary expands your ability to conceptualize, analyze, sequentialize, reason, think, define, refine, and communicate on all levels. These are all very desirable and marketable skills. They are part of your mental literacy kit.

But language isn't entirely relegated to the left brain. As you continue to develop vocabulary, it will often begin to connect with your right cortical skills. The words won't just apply to logic; they'll begin to link with your mental images, imagination, and creativity. Do you realize that if you commit yourself to learning just two new words a day, you'd enrich your vocabulary by 730 words a year?

While you can spend money on the various vocabulary courses that are available, you can also do it for free. When you hear a new word, write it down, look it up in the dictionary, use it as often as possible, 10 times a day, and it'll be yours for life. As you broaden your vocabulary, you'll significantly increase your ability to communicate, express yourself, define your thoughts, explore and define and develop your feelings, and accomplish your goals. Your brain loves receiving information. Give it to it.

LOVE AND AFFECTION

Everyone needs to feel loved and appreciated. When love is missing, the brain suffers depression and despair. It lacks the motivation to survive. Think of the terrible hurt or despair you've experienced upon the loss of a loved one. The brain, not the heart, is the center of our emotions. When it's satisfied in the area of love, you'll discover that most elements of self-management will more easily fall into place.

Love and affection come naturally at the beginning of a caring relationship. Over time, however, relationships can become strained in the midst of life's many stressors. My theory is that the source of much of the stress in nearly any close relationship comes from insufficient time and space to process information and transition from right-brain to left-brain activity and vice versa. The resulting frustration often leads to anger, which in turn results in disagreements, bitterness, and resentment. If the situation isn't resolved, the relationship faces almost certain breakup.

Allow me to use a stereotypical situation to illustrate the point. We have a husband who's been working in an office all day, predominantly using the left-brain, left-cortical skills of reading, writing, analyzing, thinking, crunching numbers, and so on. He arrives home exhausted, slumped on the doorstep, and is greeted by his wife.

In this situation, the wife is the homemaker, who's been working all day. She has been largely using her right-cortical skills. She's been involved with color, rhythm, cooking, cleaning, listening to the radio as she performs her tasks, shopping, and caring for their children. Now I'm sure you can predict what's going to happen. You've probably experienced something similar yourself.

All the husband wants to do is to rest, perhaps have a drink, settle into his favorite chair, and give his tired left brain a rest. He wants to let his mind drift. She, on the other hand, hasn't had much left-brain stimulation during the day, and she can't wait to talk, exchange ideas, and give her right brain a breather.

There they meet on the doorstep, the clash of the left and right cortical titans. She wants to talk. He wants to melt into the recliner. She applies the pressure to engage him in conversation. He's put out, thinking she's a nag and a pest who can't shut her mouth for a minute. She's furious at his lack of consideration, his collapsed mental state, his lack of appreciation, and his reluctance to engage in conversation.

By understanding how the two brains are really operating and what they really need, you can virtually eliminate this explosive situation simply by understanding the concept I call the *buffer zone*.

WHAT IS THE BUFFER ZONE?

The *buffer zone* is the time and space between the physical meeting of two parties and the commencement of their mental engagement: when two parties give each other a breather to allow their brains to settle down and come more into balance. Think of it in terms of what divers do to equalize their pressure as they swim to the surface after a deep dive. They ascend slowly, typically no faster than about 30 feet per minute, and then stop about 15 feet from the surface to check for boats. The slow ascent enables excess nitrogen, which builds up in the body during the dive, to exit safely. Failure to rid the body of excess nitrogen before surfacing can result in a painful and potentially fatal condition called the bends.

Similarly, without routinely building a buffer zone into relationship interactions, those interactions can become very painful and ultimately lead to the end of the relationship. The buffer zone gives both parties the chance to decompress. Giving each other the time and distance to decompress is a big part of what love is all about.

THE BUFFER ZONE IN A BUSINESS SETTING

The buffer zone can be modified to work in various settings, including business. In my own organization, we noticed that our secretaries were having an impossible time completing their daily work, which was getting on everyone's nerves. But it wasn't their fault. They were constantly being interrupted by the telephone and by people within the organization who needed them to do various unplanned activities.

The secretaries were frustrated, while their managers were growing increasingly impatient. As you can easily imagine, communication between the groups was not only breaking down but heating up. Just as in a personal relationship, this situation was not sustainable.

Our solution was to sit down as a team and analyze the hours spent daily on various tasks. We found that on average, the secretaries were spending two hours each day on various unplanned interruptions. Now these interruptions weren't frivolous or useless: they were very important, but they *were* unplanned. So we decided that we needed to plan for these necessary interruptions. We restructured the secretaries' days with two-hour buffer zones built into them, so now every unexpected phone call or request was built into their day and therefore expected. Interruptions were significantly curtailed, so they were no longer a source of frustration.

The result was an incredible improvement in personal relationships, which of course translated into an enormous improvement in productivity and a host of other bottom-line advantages. It was an amazing transformation of our organization.

Sleep

Discussions about the care and feeding of the brain often overlook sleep, but it is essential for optimal brain health and function. While a minority of the population can reportedly thrive on six hours of sleep per night or less, most of us require seven or eight hours of quality sleep. By *quality*, I mean restful and restorative. If you are waking up several times a night or you rise in the morning not feeling fully rested, you're not getting enough sleep or your sleep is being disrupted.

Sleep plays an important role in both brain health and function. You may not realize it, but your brain continues to function during sleep, carrying out important maintenance operations, including the following:

- **Detoxing:** Your brain has its own waste removal and recycling system, called the glymphatic system, which removes chemical waste, some of which may be toxic, and recycles other chemicals for reuse. Amyloid plaque, associated with Alzheimer's disease, is one of the key proteins that are recycled during sleep. If you're not getting enough quality sleep, toxic chemicals can begin to build up in your brain, impairing memory, cognition, and mood.

- **Repairing:** As you sleep, your brain repairs any damage it suffered over the course of the day. According to an animal study conducted at the University of Pennsylvania, extended wakefulness can damage the neurons responsible for alertness and cognition. Another study linked shortened sleep to reduced brain volume, although the researchers could not be certain whether shortened sleep caused the reduction in brain volume or the smaller brain volume impairs normal sleep.

- **Processing:** The data and stimulation your brain receives during the day are processed at night to enable your brain to catalog and make sense of it. Think of your brain's nightly processing as an enormous indexing and tagging operation, during which your internal library is organized for quick and easy retrieval of data. We know that sleep is essential for organizing and storing data because people who sleep only four to five hours a night typically perform poorly on memory tests.

- **Creating memories:** Your brain has an incredible ability to produce vivid dreams as you sleep—more vivid than the reality you awake to. These dreams are the product of the brain's attempt to make sense of all the stimulation it received during the day and its attempts to solve problems and resolve internal conflicts. Research suggests that dreams may be connected to a brain chemical called acetylcholine, which is released during sleep. In people with Alzheimer's disease, the brain cells that produce acetylcholine are destroyed, impairing their ability to dream.

Poor sleep (quantity and quality) have been linked to increased risks for stroke, cognitive aging, dementia, Parkinson's disease, and Alzheimer's disease.

Here are a few suggestions for improving your sleep quality and quantity:

- Establish a sleep routine to ensure that you're getting 7–8 hours of sleep per night. In other words, have a set bedtime.
- Sleep in a totally dark, soundproof room, or one that is as dark and silent as possible.
- No screen time (TV, computer, smartphone) for at least one hour before bed.
- Avoid caffeine, alcohol, nicotine, and other chemicals that disrupt sleep.
- Move your exercise routine to an earlier slot in the day. Don't exercise close to bedtime.

- Drink enough before bedtime so you don't wake up thirsty, but not so much that you have to wake up in the middle of the night to use the bathroom.
- Avoid eating too much before bedtime, and avoiding eating foods that are likely to disrupt your sleep, such as pizza or nachos.

Dream Hobbies

Your dream hobbies are the enjoyable, often educational activities you wanted to pursue and never have, probably because of a perceived lack of time, energy, or ability—painting, playing a musical instrument, learning another language, hang gliding, scuba diving, playing chess, sewing, welding, repairing small engines . . . whatever sparks your passion.

The very fact that you dream about these things is proof that your brain can function in those areas. The brain is saying, "I want, I need this activity." So my advice to you is this: If you have daydreams about being a painter, a sculptor, a theater actor, a world traveler, or whatever, in all probability, you are capable of doing it and doing it well, no matter what anyone has told you over the years, and no matter what you may have told yourself. So get to it. Start your dream hobby now.

Note that most dream hobbies are right-brain activities. This is apparently because most of us are left-brain dominant, and we simply don't give ourselves enough time to develop in more creative, artistic areas. However, as your brain becomes more highly developed, you begin to engage both sides for most activities. For example, sports may be mostly right-brain oriented, but there's so much else involved. Think about what a baseball player does when facing a pitcher—calculating the speed the speed and trajectory of the ball, the position of the defensive players, the speed and direction to swing the bat, and any number of other calculations. That's really complex left-brain activity. And the more skillful you become, the more you use elements from both sides of your brain. There's a balance involved that comes from doing something well.

Create a list of your top five dream hobbies:

1. ______________________________
2. ______________________________
3. ______________________________
4. ______________________________
5. ______________________________

Now choose one dream hobby from the list and commit to spending no less than one hour a week on it. If you find yourself complaining that you don't have the time, it's even more important to make the time to do it. You've got to manage yourself well enough to find time. If you think you're too old, that's nonsense. You're never too old to start taking piano lessons or to learn to hang glide, scuba dive, or skydive. We know 75-year-olds who never played a musical instrument and are now playing everything from Bach to pop. If you say you can, you will, especially when you have the right formula. When you do, it will change your life.

Revisiting the Rest-Activity Cycle

Let's return to a topic I touched on in chapter 1: the rest-activity cycle. If you were studying for some important event—such as a test, a report, or a presentation—and after 40 minutes or so were having trouble getting into the material, what would you do? You would probably get up and walk around, and maybe get a soft drink. You would do something to relax and clear your head before resuming your efforts.

However, if you were really into the material, you probably wouldn't stop to relax. You would probably forge ahead, going as long and as hard as you could. When you're on a roll, you don't stop, right? Well, you really should, and I'll tell you why: there's no doubt that your brain can continue to process the information it's ingesting, but by continuing, you lose your ability to recall the information.

Have you ever read something and understood it perfectly well, only to have quickly forgotten it? Maybe you crammed for a test and aced it but days later couldn't recall much of what you had "learned." Your short-term memory got you through the test, but that information never made it to your long-term memory.

If you want to maximize learning and recall information, you have to work within the confines of the rest-activity cycle. Depending on your ability, that means working in 20–40-minute spurts to take in information and retain it. When you give your brain scheduled breaks, you allow it to assimilate the information and prepare itself for the learning to come. Also important is taking time off—not just between reading important papers in your day-to-day routine, but to replenish yourself weekly, monthly, and yearly. You must take advantage of weekends and vacations. To manage information well, you must be able to manage your life.

Memory

Since we're on the topic of managing and assimilating information, let's talk about how to improve your memory. This is a major component of information management. If you can retain and recall facts, figures, and other key data in your personal and professional life, you'll have a leg up on everyone else. Before we get into the exercises for improving memory, I'd like to be sure you understand exactly what is going on in your brain.

Perhaps you've had that tip-of-the-tongue feeling: you know you have the information, but you just can't say it. Your brain has successfully retained the information, but you don't know how to recall it. Many scientists are certain that the brain retains virtually everything. They point to brain stimulation experiments that allow people to recall long forgotten events from dozens of years ago. The brain certainly has the capacity to remember everything. According to one study, even if the brain were fed 10 new bits of information every second for the duration of its life, it wouldn't be close to being full.

You've probably heard about near-death experiences. When people have been confronted with near death by drowning, falling off a cliff, or being hit by a truck and miraculously surviving, these are classic cases of the brain retaining virtually everything. These people report that their entire life flashed before their eyes just before they lost consciousness.

If you're thinking that's just a figure of speech ("My entire life flashed before my eyes") and that these survivors are just referring to a few major events in their lives, you would be mistaken. Upon intense questioning, they insist that in fact, yes, they experienced total recall. These reports have been gathered reliably from people of all ages, genders, and races. In each case, they have been initially reluctant to given the information, because they were embarrassed to admit what they experienced, thinking that it would make them subject to ridicule.

These individuals, not knowing that the experience was common, thought something was wrong with them, when in fact they were experiencing a moment of optimal brain function. You can recall your entire life. Although you probably never have seen your entire life flash before your eyes, you probably had a small taste of such an experience . . . or, more accurately, a smell of it. Familiar scents often bring back old memories. For example, you might be walking past a bakery and catch a whiff of homemade bread that brings back a flood of memories about your childhood, when you would spend time with your grandparents. That's called a *surprise stimulus.* A sight, a sound, a touch, or a smell can trigger an entire series of connected memories, and sometimes they are incredibly pleasant.

More evidence that our brains retain everything comes from people who can recall every day of their lives, sometimes down to the hour and minute. One Russian was studied who has what is believed to be a perfect memory. If you asked him, for example, what happened on a given day 20 years ago, he'd pause and say, "At what time?" For some reason, his brain had settled into a natural system for recalling everything in his entire life. In every other way, including the structure and functioning of his brain, he was just like you and me. If he can do it, so can you and I.

THE FIVE FACTORS THAT AFFECT RECALL

Given the vast amount of information we gather over the course of our lives, just in terms of names, places, dates, and events, it's easy to believe that we've been exposed to far too much information for us to recall *everything*. But that's not the case. We don't recall information simply because we don't know *how* to recall it. The better your filing system, the easier it is to obtain the data you need. Here are the real factors that affect your recall:

- **Primacy:** You will usually recall the beginning of events (*primacy* meaning *first*) more than the middle of events. You'll recall the first time you did something—for example, when you learned to ride a bicycle—more than the times you repeated that action. You'll have the opportunity to see that for yourself when you get into the exercises later in this chapter.

- **Recency:** All things being equal, you tend to remember events that just happened—the most recent events. In other words, you'll remember what you did yesterday better than the day before that, and what you did two days ago better than what you did three days ago. This is true even in old age, where people can recall events from their very early childhood and from very recent times but lose much of their middle years. It's not so much age; it's simply the pattern of recall.

- **Linking:** You recall things that you connect or associate with other things; for example, if somebody tells you something that evokes a certain image, you're much more likely to be able to recall what they said because it's linked to that image. The linkage may be between words, phrases, images, events, smells, sounds, feelings, or any combination of these. Linkage is a critical aspect of developing a superior memory.

- **Outstandliness:** In psychological jargon (for any of you who want to use that language), what I refer to as "outstandliness" is known as the Von Restorff effect. Among members of my generation, nearly everyone can recall where they were and what they were doing, for example, when President John F. Kennedy was shot and killed. For younger people, it could be the attack on the Twin Towers on 9/11/2001. We have an automatic system for recalling things or events that stand out from all others.

- **Willingness to review:** Many physiologists now claim that the brain pattern is strengthened by repeating the memory pattern. In other words, your thought affects your physical internal brain structure. Things that you review will be more firmly lodged in your brain than things that you skim over only once. In other words, if you want to remember something, review it.

Now that you know how the brain recalls information, you need to be able to apply that knowledge. Given the five factors that affect recall, what's wrong with studying four hours without a break? When you study nonstop for a long period of time, you're giving yourself only one shot at primacy and recency. You'll be able to recall the material at the beginning and at the end, with a big sag in the middle. By taking frequent breaks, you increase your opportunities for primacy and recency, and if you add some review into your study routine, you can significantly increase your recall potential.

Primacy, recency, and review all reinforce the value of the rest-activity cycle. By organizing your learning periods into 20–40-minute sessions (depending on the difficulty of the subject matter, your experience with the material, and your level of interest), you'll not only be giving your recall a big boost, you're going to feel considerably more rested. You've given your mind a chance to rest and sort out the information.

CHUNKING

Chunking involves breaking down information and activities into easier-to-manage segments. You break down study and practice sessions to 20–40 minutes with 2- to 5-minute breaks between them. You break down subjects into topics and subtopics. You break down paragraphs into lists of no more than seven items each. Chunking is one of the keys to organizing yourself: you divide, group, and organize items into packets that are easier to manage and recall.

Many of us make the mistake of trying to remember everything on a grand scale. If you think in chunks, if you can categorize things into compartments, you'll be able to remember and manage them more effectively.

If you pride yourself on being a multitasker, that's fine. Having the ability to juggle a number of responsibilities at once is a valuable skill, but it depends on the number of balls you're juggling. Research seems to indicate that we fail not because we don't have ability, but because we don't accommodate the brain's natural organizing and chunking processes.

Focus is a key to success in any area of life. You need to know exactly what you can do and what you are capable of and work to create success within that framework. If you know that the brain works best dealing with just so many activities at a time (seven is a maximum for most people), then work with no more than that number of ideas or concepts at a time. This will help you gain a clear picture of where you are and where you're heading.

MNEMONIC DEVICES

People have come up with all sorts of clever ways to remember and recall information and key concepts in grade school, high school, and beyond. These mnemonic devices can be broken down into the following categories:

- Imagery
- Acronyms
- Acrostics
- Rhymes
- Chunking and organizing
- Models

IMAGERY

Imagery is one of the most effective mnemonic devices, because it enables you to link a word, phrase, concept, or another image with a vivid image in your mind. For example, if you meet someone named Sandy, you might picture her on a beach surrounded by sand. To recall the difference between the words *principle* and *principal*, remind yourself that the principal can be your pal, but a principle cannot.

ACRONYMS

An acronym is one or more words formed by using the first letters of the items you want to remember; for example, you may recall that the colors in a rainbow spell *Roy G. Biv*: red, orange, yellow, green, blue, indigo, and violet. You can remember the names of the Great Lakes with the acronym HOMES: Huron, Ontario, Michigan, Erie, and Superior. You treat a sprained limb with RICE: rest, ice, compression, elevation.

ACROSTICS

An acrostic is similar to an acronym, but it uses the first letters of the items you want to remember to create a sentence. For example, to remember the names of the planets in the solar system in the order they are starting nearest to the sun, you can use the acrostic *My Very Educated Mother Just Served Us Nine Pies*: Mercury, Venus, Earth, Mars, Jupiter, Saturn, Uranus, Neptune, and Pluto (when Pluto is counted as a planet).

To recall the order of operations in math, you may have been taught the following acrostic: *Please Excuse My Dear Aunt Sally*—parentheses, multiplication, division, addition, and subtraction. In music class, you may have been taught the order of the treble notes as *Every Good Boy Does Fine* (EGBDF) and the order of the bass notes as *All Cows Eat Grass* (ACEG). In geometry, you may have been taught the acrostic *soh cah toa* to remember the formulas for sine, cosine, and tangent:

- Sin = Opposite/hypotenuse
- Cosine = Adjacent/hypotenuse
- Tangent = Opposite/Adjacent

RHYMES

As human beings, we're geared to remember and recall rhymes much more easily than prose. In fact, you probably still remember these classic rhymes from your school days:

- Thirty days hath September, April, June, and November. All the rest have 31, excepting February alone, which has but 28 days clear and 29 in each leap year.
- In fourteen hundred and ninety-two, Columbus sailed the ocean blue.
- I before E, except after C or when sounding like A, as in *neighbor* and *weigh*.
- Acid to water, that's what you oughter.
- Rainbow in the morning, travelers take warning. Rainbow at night, travelers' delight.

CHUNKING AND ORGANIZING

With chunking and organizing, you break down information into smaller units that are easier to remember and recall, and you group like items together. Here are a few examples:

- Phone numbers, credit card numbers, and Social Security numbers are all broken down into groups of two to four numbers. For example, the phone number 8777325254 is much easier to recall in its standard form: (877) 732-5254.

- To recall items on a grocery list, consider grouping them by item type or section of the store—for example, dairy (milk, cheese, eggs); produce (apples, bananas, and lettuce); canned goods (canned fruits, veggies, and beans); frozen foods (prepared dinners, ice cream, frozen juices); and so on.
- Use colors to keep track of priority items, such as green for priority projects, yellow for projects that are next on your list, and red for projects that are on hold for some reason.

MODELS

Models are a form of imagery. Think of them more in terms of the images or charts you're likely to see on an infographic, such as an organizational of flow chart, a lifecycle diagram, or a pyramid, such as the food pyramid, which serves as a visual guide for how much to eat from each food group: bread, cereal, rice, and pasta; vegetables; fruits; dairy; protein; and fats, oils, and sweets.

Feats of Memory

Since the time of the ancient Greeks, people have amazed others with remarkable feats of memory. Some have been able to remember hundreds of items backwards, forwards, or in any order whatsoever. It could be dates and numbers, names and faces, entire bodies of knowledge in a given subject area. Others have memorized long stories, passing them down orally from one generation to the next.

In more modern times, memory champions have demonstrated their skills by memorizing several decks of playing cards shuffled to place the fifty-two cards in each deck in a random order. They can then recite the names of the cards in perfect order. Until recently, the use of mnemonics memory techniques were generally scorned as magic tricks. But no sleight of hand is at work here. As other researchers and I have discovered, these people were actually using techniques that mirror the way the brain functions.

The preliminary mnemonic techniques that you are about to learn are based on the way your brain functions. They are not tricks, and they're now being taught in leading colleges, universities, and major corporations around the world. These techniques are designed for the way your brain works. They enable you to develop your mental muscle.

Memory Exercises

The brain turns in an amazing performance at a subconscious level, managing and coordinating all the physiological activities required to keep us alive. But we also need to function at a conscious level. We need to perform productively at school or at work, fulfill our domestic and civil responsibilities, carry on relationships, and engage in recreational activities. To improve your performance at a conscious level, start by improving your memory. Work with me to improve your conscious self, and you'll be well on the way to attaining your goals.

You've learned some of the reasons why we can and can't recall information. Now we'll start to develop your memory power so that every bit of information will be at your fingertips when you need it.

Hyperthymesia

Hyperthymesia or highly superior autobiographical memory (HSAM) is a condition that enables a person to remember and recall an abnormally large number of their life experiences in vivid detail. The condition affects a very small percentage of the world population—only about 30–60 individuals in the world at any given time.

On YouTube, you can still access the *60 Minutes* piece about actress Marilu Henner and two other women who have been tested and confirmed to have hyperthymesia. For two of the women, the condition has been a blessing, but for the third, it has been a curse: her mind is cluttered with memories and information that impair her ability to focus on daily activities.

During the piece, Henner's son asks her, "What day was Valentine's Day in 1979?" and she immediately replies, "It was a Wednesday," which is correct. He complains that she never explains how she does it, and she replies, "I don't do it. I just see it." She can recall nearly every day of her life, including many of the important world events that occurred on those days.

Researchers scanned the brains of these individuals and found something out of the ordinary—an enlarged temporal lobe and a caudate nucleus up to seven times the normal size. The caudate nucleus is associated primarily with procedural memory and is closely linked to obsessive-compulsive disorder. While none of the women in the *60 Minutes* piece was diagnosed as having obsessive-compulsive disorder, all were highly organized.

EXERCISE 1

Below is a list of words (don't look at them yet!). When you have finished reading these instructions, read through the list at your normal speed one word at a time from start to finish. Don't back up and reread any words. There are too many words in the entire list for you to remember them all, so the task is simply to remember as many as you can. When you're finished reading through the list, answer the questions that follow.

pay	which	then
head	the	of
turn	will	they
now	once	actual
fee	and	of
field	more	and
the	clearly	case
of	Leonardo da Vinci	the
left	together	repeat
and	inch	same
to	and	other
of	the	

Now, without looking back, answer the questions.

1. How many of the first six words in the list can you remember?

 ____________ ____________ ____________

 ____________ ____________ ____________

2. How many of the last six words can you remember?

 ____________ ____________ ____________

 ____________ ____________ ____________

3. Can you remember any word which appeared more than once?

 ____

4. Can you remember any word or group of words that was outstandingly different from all the others?

 ____

5. Can you remember any other words?

 __________ __________ __________

Well, how did you do? Don't be embarrassed if you remembered only a few words. This exercise is intended to help you understand (a) that you can improve your memory and (b) probably more importantly, that you can understand *how* you remembered. If you remember what I said about primacy, you tend to recall more from the beginning than the middle. You probably remembered one or two words at the beginning and maybe a few more at the end, thanks to recency—retaining more of what you recently learned.

Were you able to recall any words that appeared more than once? If you did, you benefited from linking or review. The repetition of words was enough to stamp them into your memory. That proves that it's easier to recall things that are somehow connected rather than disconnected.

Did you find any word or group of words that was remarkably different? Chances are good that *Leonardo da Vinci* jumped off the page and made quite an impression. Thanks to the name's outstandliness, you were able to recall it almost effortlessly.

How many other words you remembered doesn't matter. However good your memory is now, you'll make it better within the next 15 minutes.

EXERCISE 2

In this exercise, associate each number with a specific item. One is *table*, two is *feather*, three is *cat*, and so on. As with Exercise 1, read each item once, covering the ones read with a card as you progress down the list. The purpose of this is to remember the number-word pairs:

4 leaf

9 shirt

1 table

6 orange

10 poker

5 student

8 pencil

3 cat

7 car

2 feather

Now cover this list and fill in the answers in the order requested below.

Here are the numbers 1 to 10. Fill in next to each number the word that appeared next to it. The numbers are listed in a different order intentionally. Do not refer back to the list until you have filled in as many as you can.

1	______________	7	______________
5	______________	4	______________
3	______________	6	______________
8	______________	10	______________
9	______________	2	______________

Score: ______________

To remember these, it is necessary to have a system that enables the use of the associative and linking power of memory to connect them with their proper number.

The best system for this is the Number-Rhyme System, in which each number has a rhyming word connected to it.

The rhyming words are:

1. bun
2. shoe
3. tree
4. door
5. hive
6. sticks
7. heaven
8. skate
9. vine
10. hen

The purpose of this exercise is to improve your power to recall. Just work with me for a bit. In order to remember the 10-word list you just tried to recall (or anything, for that matter), you need to link the words with those rhyming words connected to the numbers. In a short period of time, you'll be able to recall instantly the rhyming word with the rhyming number; with it, you'll also recall the image of whatever you have to remember.

EXERCISE 3

As a review, check your improving memory once again. In the spaces below write the rhyming key word for the Number-Rhyme System, and next to it the words used earlier in the chapter to illustrate the system.

I want you to imagine what I'm going to tell you. I want you to get as clear a picture as possible in your mind's eye. Use your senses to experience whatever I describe. Feel it, hear it, smell it, or experience it in any way that works for you. As an example, take a moment to close your eyes now and think of what you had for dinner yesterday. Engage as many senses as possible—sight, smell, taste, touch, and hearing.

In the future, you're going to create your own imagery, because that's what will work best for you. Now let's get back to work. I'm going to present the number and the word that rhymes with the number.

Then I'm going to give you the word I want you to remember from the same group of words presented in Exercise 2. Here we go:

1. One, bun. The word to remember is *table*. Think of a giant bun sitting on a spindly table whose legs are starting to collapse under the bun's weight of the bun. Smell the bun; inhale that wonderful bakery-fresh aroma. Taste it.
2. Two, shoe. The word to remember is *feather*. Think about getting dressed for work, but as you try to put on one of your shoes, you find you can't. There's a giant feather growing on the inside; every time you put your foot in

the shoe, the feather tickles you and makes you laugh. Just think about that feather inside your shoe, tickling the bottom of your foot.

3. Three, tree. The word is *cat*. If you own a cat or know someone with a cat, think of that cat stuck way at the top of a tree, desperately holding onto skinny branches and meowing at the top of its lungs. Can you sense that cat's anguish? It's swaying back and forth.
4. Four, door. Your word is *leaf.* Imagine that the door to your bedroom is really a giant leaf and every time you open or close the door, it makes a crunchy sound, just as dried leaves do in autumn.
5. Five, hive. The word to remember is *student*. Think about a student sitting at a desk. She's wearing a black and yellow striped shirt, and she's buzzing happily while she writes in her notebook with a pen filled with honey.
6. Six, sticks. The word is *orange*. Think of a giant orange, the biggest one you've ever seen—as big as a beach ball. Some enormous sticks puncture the orange, squirting the juice all over you. Smell the juice. Feel it oozing over your body.
7. Seven, heaven. The word to remember is *car.* Imagine that every angel in heaven is sitting not on a cloud, but on a car, and the car each one is sitting on is the one you think is the most heavenly car imaginable, the one you can envision yourself driving.
8. Eight, skate. Your word is *pencil*. Think of yourself on roller skates with colored pencils attached to them. As you roll down the street, the pencils make fantastic multicolored designs behind you.

9. Nine, vine. The word is *shirt*. Do you remember the giant vine in "Jack and the Beanstalk"? Instead of leaves flapping on the side of the vine, imagine the beautiful pastel-colored shirts are attached to the vine. When the wind blows, the shirts create a beautiful display on the vine.
10. Ten, hen. Your word is *poker*. Think of a group of men playing poker. Whenever one runs out of money, he squeezes the hen, and more poker chips come out.

OK. Open your eyes. Let's see how much the number rhyme system helped you recall your key words. Fill in the blank spaces below with both the number rhyme word and your key word.

	Rhyming Key Words	Word Connected
1.	____________	____________
2.	____________	____________
3.	____________	____________
4.	____________	____________
5.	____________	____________
6.	____________	____________
7.	____________	____________
8.	____________	____________
9.	____________	____________
10.	____________	____________

Did you show any improvement? Any improvement you notice can be attributed to the images you formed in your mind. To become perfect, all it takes is practice. As you continue to develop your recall, be certain that the rhyming word and the word you want to remember are tightly linked. The examples that I provided may not suit your own particular linking style. In the following section, I describe the essential ingredients of the mental images you form that will make them work most effectively for you.

The S-M-A-S-H-I-N' S-C-O-P-E of Memory

Whenever you're using rhyming words and images to memorize a list of words, be sure the rhyming word and the word you're committing to memory are totally and securely linked. To establish a strong link, create mental images that meet one or more of the following criteria (the more, the better).

SYNESTHESIA/SENSUALITY

Synesthesia refers to the blending of the senses. The great "natural" memorizers (mnemonists) have acute senses, and they blend sensory information from all their five senses to enhance recall. As you take in sensory data over the course of a day, train yourself to be more conscious of it. Simply shift your focus from one sense to another:

- Sight
- Hearing
- Smell
- Taste
- Touch
- Kinesthesia: your awareness of bodily position and movement in space

MOVEMENT AND DIMENSION

In any mnemonic image, movement and dimensions add a giant range of possibilities for your brain to "link in" and thus remember. As you imagine objects, make them three-dimensional and put them in motion.

ASSOCIATION

Whatever you wish to memorize, make sure you associate or link it to something that's a part of your regular mental environment: for example, three = tree. You could use any rhyming word instead, such as three = brie (the French cheese), but only if brie is a part of your everyday experience or environment, or maybe your favorite cheese! Link to what you know and are familiar with.

SEXUALITY

Now this is something we all have perfect memories for. Be delighted to use sexual images to make your linkages and associations. Remember, these are private images. No one can get in there and have a look. They're not for anyone else to experience and know; they're yours. I don't use sexual images in the book for obvious reasons, but if the images in your mind help you remember, by all means use them. Sexuality provides a rich source of vivid imagery.

HUMOR

Have fun with your memory. The funnier, more ridiculous, absurd, and surreal you make your images, the more outstandingly memorable they will be. Salvador Dali, the surrealist painter, once said, "My paintings are photographs painted by hand of the irrational made concrete." In many instances, they were the paintings of perfectly held memories of his day-time and nighttime dreams.

IMAGINATION

The I in SMASHIN' SCOPE stands for *imagination*. Einstein said, "Imagination is more important than knowledge. For knowledge is limited, whereas imagination embraces the entire world, stimulating progress, giving birth to evolution." The more you use your imagination, the better you will become at imagining. And the better you get, the more astonishing your memory powers will become.

NUMBER

The N in SMASHIN' SCOPE is for *number*. Now you're getting really good at numbers. By numbering items, you give them an order and sequence that helps you remember them again, using your left- and right-brain cortical skills.

That completes the word SMASHIN' and brings us to the second word, which is SCOPE.

SYMBOLISM

The S in SCOPE stands for *symbolism*. Normal symbols are fine, but the more meaningful a symbol is to you and the more exaggerated and imaginative and colorful, the greater the chance will be that you'll recall whatever you've associated with it. Substituting a more meaningful image for a more normal or boring image increases the probability of recall. So make it fantastic. You may also use traditional symbols (for example, a stop sign or light bulb).

COLOR

The C stands for *color.* The more vivid they are, the more memorable your mental images become. The more memorable they become, the better your memory becomes and the more colorful your life becomes. Where appropriate, and whenever possible, use the full range of the rainbow to make your ideas more colorful and therefore more memorable.

ORDER/SEQUENCE

The O in SCOPE is for *order*: sequence. When you fully develop this process, you increase the brain's ability to access information like a well-ordered library. In combination with the other principles, order/sequence allows for much more immediate reference and increases the brain's possibilities for "random access."

POSITIVITY

The P is for *positivity.* Keep your images positive and pleasant so that your brain will want to come back to them. If you deal with negative images, even in conjunction with all the other principles, you may find the brain blocking their recall because it finds them unpleasant or offensive and doesn't want to remember them.

EXAGGERATION

The final letter in SMASHIN' SCOPE is the E, which stands for *exaggeration.* Make everything exaggerated. It takes us back to that principle of using the outstanding. Don't just put a bun on the table: make it a giant, delicious bun oozing all the things you love to eat, so big that it makes the table collapse, and you hear it collapsing and you taste it and feel it all over. Get involved. Don't just put a stick through a regular orange; make that orange as big as a planet, stick a gigantic stick through it, and feel the juices, taste it, get it all over you. Get involved, make things big and delightful, and you'll remember them incredibly well.

If you apply the SMASHIN' SCOPE, your memory will expand its dimensions infinitely. In all your images, exaggerate size, shape, and sound.

Many memory systems are available that can help you develop your memory to the point where you'll never have to take notes at a meeting or you can answer any question any time without a note. These aren't tricks; these are information management systems that can be of immense value in putting every bit of information at your fingertips. They can serve as a tremendous aid in problem solving and decision making. Practice the number rhyme system, along with SMASHIN' SCOPE, and add to your knowledge with other memory builders.

What You Can Do with Fully Developed Memory Skills

People who have excellent memories have the potential to excel at school, at work, and in their personal lives, because they have information at their fingertips. Let me tell you an interesting little story to illustrate my point. This is a story about a class of 14-year-old students in Sweden, who were given what their teacher described to them as an impossible task. He instructed them to simply do as best they could.

They were asked to memorize as many of the countries and capitals of the world as possible, a number well over 300. One of the boys in the class came home and complained to his dad about what he felt was an unfair and unrealistic assignment. His father, however, was a senior manager at a large company, where he had taken one of my memory courses. He told his son, "Look, I can show you how to do this." The child was surprised to hear his father tell him that the assignment wasn't really difficult at all.

The boy returned to school and took the test. A few weeks later, the father received a call from the principal of the school advising him that his son was found to have cheated in the geography test. The dad came to the school to discuss the issue. The only proof that the school had to implicate his son was the fact that he had done so well on the test. The second highest score was 123. His son scored over 300! The teacher concluded that such a high score was impossible, and therefore the boy must have cheated.

The boy could have taught the headmaster a lesson or two, and that is exactly what he did . . . and more. When they realized that the boy had actually applied his amazing brain power to the task, they asked him to teach my memory techniques to all his classmates. Together, they raised the learning curve of that school.

CHAPTER 5

Getting to Know Yourself:

Self-Exploration

Our greatest human adventure is the evolution of consciousness. We are in this life to enlarge the soul, liberate the spirit, and light up the brain.

—Tom Robbins

We're about to enter an entirely new phase of our brain power program. In preparation, I'd like you to do three exercises. They will help you analyze your life and determine where you are in your personal development and where you would like to be. They will also indicate the amount of time left in your life to enjoy the fruits of meaningful change.

This is the beginning of learning how to use your brain power to create a life that is more in tune with who you are and what you really want for yourself, your family, and other loved ones. By working through the next group of exercises, you will find that your future will never be one in which external circumstances or other people dominate you. *You* will be in control.

The process of self-examination is your next step to unleashing the extraordinary range of mental skills you're starting to become aware of—your full range of mentally literate skills. You're going to be answering questions about yourself, who you are, what you want to accomplish, and what you want to become. The long-term benefits will be, I promise, dramatic. At the beginning, you'll find you will easily be able to overcome any initial hesitancy because change, as we now know, can be accomplished with ease.

Here's an example. Most people who smoke cigarettes do so to conceal feelings of social inadequacy and other frustrations in their lives. They don't know what to do with their hands, and they need something to keep them occupied during embarrassing pauses and other social situations. As they continue the habit over the years, it becomes an involved and complex ritual. For example, a woman acquaintance of mine who had smoked for years finally stopped. When she did, she found that her feelings of social inadequacy were magnified tenfold; she became frantic because she no longer had her prop. This apparent setback was actually a positive event, because it allowed her to take a clear look at herself and focus on the underlying causes behind her emotions, so she could approach the problem head-on rather than avoiding it for the rest of her life. She has stopped smoking, and she has been able to shed her feelings of social inadequacy, which she had masked with that unhealthy prop.

Exercise 1: Journaling

One of the best ways to discover who you really are is through journaling. You can keep a daily journal of your thoughts and experiences in a blank tablet or electronic document, or you can use specialized journals available through bookstores. You can also employ smartphone apps that include prompts to encourage self-exploration. To get started, respond to the following 10 prompts:

1. Name your five top strengths:
   - ______________________
   - ______________________
   - ______________________
   - ______________________
   - ______________________

2. Name five weaknesses you would like to overcome:
   - ______________________
   - ______________________
   - ______________________
   - ______________________
   - ______________________

3. Describe a significant challenge you faced in your life and explain how you overcame it:

   ____

   ____

   ____

   ____

   ____

4. If you could accomplish anything, what would it be?

____

____

____

____

____

5. If you knew you couldn't fail, what bold and daring activity would you try?

____

____

____

____

____

6. What do you enjoy doing most?

____

____

____

____

7. List your values in order of importance (for example, honesty, loyalty, independence):

- ____
- ____
- ____
- ____
- ____

8. What qualities do you value most in a friend?
    - ______
    - ______
    - ______
    - ______
    - ______

9. What are your biggest fears?
    - ______
    - ______
    - ______
    - ______
    - ______

10. Describe the five best moments of your life.
    - ______
    - ______
    - ______
    - ______
    - ______

Self-exploration through adventure

One of the best ways to get to know who you truly are is through adventure—trying something outside your comfort zone, something that strikes you as challenging or even a little scary. Think about it. You never really know how smart you are, how strong you are, or how resilient you are until your intelligence, strength, and resiliency are pushed to their limits. In addition, adventure usually introduces you to something larger than yourself and the confines of where you live, which enables you to view yourself from a different perspective.

Here are a few options for pushing yourself outside your comfort zone:

- Attend a venue or event that's unfamiliar to you.
- Travel alone to a country where you don't speak the language.
- Do something that scares you a little, such as skydiving or rock climbing.
- Take a course on an unfamiliar subject.
- Tackle a challenging problem head-on.
- Connect with someone new.

Exercise 2: Writing Your Own Obituary

Imagine that you are the obituary columnist for your local newspaper, and that just as you are reading this section of the book, an email arrives informing you of your own death, that very morning! Included in the email message is a brief note from the editor requesting an immediate obituary about you for next morning's edition.

In the space provided, write your own "real" obituary, to date. Take as much time as you need. In fact, don't move on to the next chapter until you complete it. I really want you to think about it. Take an hour, a few hours, a day, or however much time you need to really think about what you've accomplished so far in your life—all aspects of your life. You don't have to confine yourself to the space provided in this book.

OBITUARY

(use additional paper as necessary)

Note: If you find that your obituary was satisfactory to you, you have a solid foundation on which to build your future. If you find that you would like to have contributed more to the human race than you have to date, the remaining exercises will help you to make appropriate adjustments and set goals that are more in tune with your true ideals.

Today is the first day of the rest of your life. In that spirit, I want you to do one other thing before moving on to the next chapter. I want you to compute your personal life expectancy. This is a key for understanding not only the duration of your life, but the quality of your life. In the next chapter, you will learn why you did these exercises and how my system will help you take control of your life and build the future you desire. You're really moving into the heart of this program on unleashing the full range of your mental powers.

Exercise 3: Your Personal Life Expectancy Questionnaire

The balancing of your left-brain and right-brain activities, the food you eat, the amount of exercise you take, the managing of your personal relationships, the degrees of stress to which you expose yourself, your health maintenance habits, and your overall health management affect both the quality of the life you live and the duration of your life. Now that you have examined the person you are—the person you have become over the course of your life—complete the following questionnaire to determine roughly how much time you have left to become the person you want to be and to build the life you envision.

ESTIMATING YOUR LIFE EXPECTANCY

A growing number of psychologists and physiologists believe that many of us have at birth the capacity to live to between 85 and 120. After many years of research on longevity, it is possible to give general guidelines, in the form of a questionnaire, which enable you to work out roughly how long you have to live.

Start by looking up your age in the Long-Life Table below, which will enable you to estimate your basic life expectancy based on figures produced by insurance actuaries. Then, in answering the questions on the following page, add to or subtract from this number. See the section "Longevity Questionnaire Tabulation" for a table you can fill out to record your longevity data and do the math.

Remember one thing: women can expect to live roughly three years longer than men. Women, therefore, should add three years to start with.

THE LONG-LIFE TABLE

To estimate your life expectancy, start with your basic life expectancy. Look up your age in the table on the next page, circle the basic life expectancy corresponding to your age, and record it in the space provided in the Longevity Questionnaire Tabulation table. Then answer the questions on the following page and note the amount of time to add or subtract for each answer in the Longevity Questionnaire Tabulation table.

Basic Life Expectancy Chart

Present Age	Est. life expectancy	Present Age	Est. life expectancy	Present Age	Est. life expectancy
15	70.7	39	72.4	63	77.3
16	70.8	40	72.5	64	77.7
17	70.8	41	72.6	65	78.1
18	70.8	42	72.7	66	78.4
19	70.9	43	72.8	67	78.9
20	71.1	44	72.9	68	79.3
21	71.1	45	73.0	69	79.7
22	71.2	46	73.2	70	80.2
23	71.3	47	73.3	71	80.7
24	71.3	48	73.5	72	81.2
25	71.4	49	73.6	73	81.7
26	71.5	50	73.8	74	82.2
27	71.6	51	74.0	75	82.8
28	71.6	52	74.2	76	83.3
29	71.7	53	74.4	77	83.9
30	71.8	54	74.7	78	84.5
31	71.8	55	74.9	79	85.1
32	71.9	56	75.1	80	85.7
33	72.0	57	75.4	81	86.3
34	72.0	58	75.5	82	87.0
35	72.1	59	76.0	83	87.6
36	72.2	60	76.3	84	88.2
37	72.2	61	76.6		
38	72.3	62	77.0		

LONGEVITY QUESTIONS

1. Add one year for each of your grandparents who lived to be eighty or more. Add half a year for each one who topped seventy.

2. Add four years if your mother lived beyond eighty, and two if your father did so.
3. Add two years if your intelligence is above average.
4. Take off 12 years if you smoke more than 40 cigarettes a day; subtract seven years if you smoke 20 to 40; subtract two years if you smoke fewer than 20.
5. If you enjoy sex once or twice and week, add two years.
6. If you have an annual checkup (a thorough one), add two years.
7. If you are overweight, take off two years.
8. If you sleep more than 10 hours every night, or less than five, take off two years.
9. Alcohol consumption: One whiskey, half a bottle of wine, or four glasses of beer three times a week, count as moderate; add three years. Light drinkers—that is, those who drink very little per week—add only one and a half years. If you don't drink at all, don't add or subtract anything. Heavy drinkers or alcoholics, subtract eight years.
10. Exercise. Three times a week—running, cycling, swimming, brisk walking, dancing, or skating—add three years. Don't count weekend walks or other occasional light exercise.
11. Do you prefer simple, plain foods, vegetables, and fruit, to richer, meatier, fatty foods? If you can say yes honestly and can stop eating before you are full, add one year.
12. If you are frequently ill, subtract five years.
13. Education. If you did postgraduate work at university, add three years. For an ordinary bachelor's degree, add two. Up to grade 12, add one; grade 10 and below, none.

14. Jobs. If you are a professional person, add one and a half years; technical, managerial, administrative, and agricultural workers add one year; proprietors, clerk, and sales staff add nothing; semiskilled workers take off half a year; laborers subtract four years. If, however, you're not a laborer but your job involves a lot of physical work, add two years. If it is a desk job, subtract two years.
15. If you live in a city or town or have done so for most of your life, subtract one year. Add a year if most of your time has been spent in the countryside.
16. If you have one or two close friends with whom you can confide everything, add a year.
17. If you are regularly able to rest and enjoy taking it easy, add two years.

LONGEVITY QUESTIONNAIRE TABULATION

Basic Life Expectancy (from table): ________________

	Years Subtracted	Years Added
1.		
2.		
3.		
4.		
5.		
6.		
7.		
8.		
9.		
10.		
11.		

	Years Subtracted	Years Added
12.		
13.		
14.		
15.		
16.		
17.		

Totals: Basic Life Expectancy: __________

– Years Subtracted: __________

+ Years Added: __________

= Actual Life Expectancy: __________

Fourteen Tips for Achieving a Long, Happy, and Healthy Life

Your life expectancy is not predestined. From the 1500s to the 1800s, life expectancy in Europe hovered around thirty to forty years of age. Thanks to improved health care, sanitation, immunizations, access to clean running water, and better nutrition, life expectancy in most industrial nations is around seventy-five years of age.

You can take steps to push that number even higher. Here's a list of 14 steps you can take to increase your life expectancy:

1. Improve your intelligence by using your left and right brains together and increasing your vocabulary. Imagining numbers and computations instead of verbalizing them is one way to use your left and right brains

together. Improved vocabulary also engages left-brain and right-brain thinking.

2. If you smoke or vape tobacco products, stop or cut down immediately to improve your physical and mental health. Remember that oxygen is essential for brain health and function. Smoking not only reduces the flow of oxygen to the brain but also exposes it to toxins.
3. Maintain a healthy weight, typically the low to medium end of the average weight for your height, but don't obsess over it. If you're muscular, weighing more toward the high end of that range is fine. You can find charts for healthy weights based on height online.
4. Eat mostly whole foods—a mostly plant-based diet—to optimize digestion, circulation, nutrition, and cognitive function. Avoid sugar and artificial sweeteners, sweets (including sweetened beverages), and simple carbohydrates (such as those in chips, breads, bakery items, and crackers).
5. Get seven to eight hours of quality sleep per night. If you're waking up not feeling rested, you're not getting enough sleep, or the sleep you're getting is not restful and restorative.
6. Exercise at least 30 minutes most days of the week, alternating between cardiovascular workouts and weight or resistance training.
7. Include time for lovemaking. Because lovemaking seems like a natural part of intimate sexual relationships, many people neglect it; they don't make a conscious effort to include it in their daily or weekly activities. Building sex into your daily or weekly schedule may seem overly

clinical, but lovemaking is a unique activity that ideally combines love, affection, and exercise.

8. Drink mostly pure water. Unsweetened coffee and tea are fine. Alcohol in moderation is OK. Moderate drinking means one shot of hard liquor, a half bottle of wine, or four glasses of beer up to four times a week.
9. Have regular medical checkups, and work with your doctor to use the fewest medications at the lowest doses necessary to treat any medical conditions. Review your medications in close consultation with your prescribers every 6–12 months.
10. Commit to a life of continuous education. Always be in the process of learning something—study a subject, learn to play a musical instrument, study another language, practice your favorite hobbies, start a business. The goal here is to constantly challenge your brain.
11. Make sure that your job includes a reasonable balance of mental and physical activity.
12. Spend time in nature (it's good for the right brain as well as the lungs!).
13. Develop at least one close friendship, and maintain an active social life.
14. Relax, engage in playful activities, and enjoy yourself.

CHAPTER 6

Mastering Life Management

The best way to predict your future is to create it.

—ABRAHAM LINCOLN

The point of the self-exploration you conducted in the previous chapter is to give you a sense of yourself, to determine whether you have a solid foundation upon which to build your future, and to see how much time you probably have remaining in your life to achieve your goals.

If you're completely happy with where you are and what you've accomplished in life so far, not just in work, but in terms of relationships, love, and your contributions to society, you can probably skip this chapter. Nevertheless, you'll probably want to continue reading just to make sure you don't miss something that can help even more.

At this point in your brain power self-development program, you are aware that your human information processing skills are better and faster than those of any computer, and you know that they can improve far beyond where they are now. You've learned the basics about your ability to deal with logic, numbers, order, sequencing, and the other bits of intelligence stored in your cortex. By now you understand your dream hobbies and what they say about your capabilities. You're on the road to dramatically improving your memory skills, and you're about to get further into life management—a critical skill for information management.

Chances are good that you're not totally satisfied with your life: you haven't made the kind of contribution you expected or desired. Based on your answers to the questions in the life expectancy questionnaire, and your calculations, you have a general idea of how long you have left to turn your life around and make some serious progress toward self-fulfillment. Maybe you have a window of 10, 20, or even 50 years. Even on the low end, that is a considerable amount of time to make a change. In this chapter, you begin the process by taking control of your life.

Organizing Your Life

You've heard the expression about a person biting off more than she can chew. Well, that's what most of us do. We take on too much and try too hard, which leaves us feeling overworked and overwhelmed, and we end up undermining our own performance. To have any hope of achieving a balanced life and self-fulfillment, we need to break down our goals, responsibilities, and desires into more easily managed divisions. I touched on the topic of chunking in chapter 4. Just as you can improve your memory with chunking, you can improve your ability to manage your life by breaking it down into categories and prioritizing those categories.

For optimum recall, we chunk information into no more than seven items at a time. Any more than seven, and recall performance starts to suffer. If you try to take on too much, you're setting yourself up for failure. Similarly, to gain control of your life, I recommend dividing your interests and activities into a maximum of seven major divisions. This enables you to leverage the natural tendencies and abilities of the brain. Organization gives you freedom. Disorganization imprisons you.

You can use any divisions that are most relevant to your life, goals, and desires. Everyone's divisions will be unique—a reflection of who they are and the life they desire. You may have far more than seven divisions, but you need to organize them into no more than seven *major* divisions. Here are some divisions that people have used in the past:

- Quality of life
- Home
- Family
- Work
- Business
- Hobbies
- Relaxation
- Creativity
- Travel
- Play
- Self-improvement
- Reading and study
- Community
- Pets
- Environment
- Writing
- Entertainment
- Culture
- Social life
- Learning
- Law
- Household
- Emotional stability
- Sensuality and sexuality
- Finance
- Accounts
- Nutrition and food
- Sports and recreational activities

As you proceed through this chapter, you will have a lot to think about. You've laid out your life expectancy and have begun to think about organizing your life into major and minor divisions. No matter how young or old you are now, you have only so much time on this planet. What you do with those years is totally up to you. By organizing your life, by taking just a few more hours to understand the complete information management and information processing systems in this chapter, you will give yourself the key to living life to its fullest.

Exercise 1: Ranking Your Seven Divisions

In the space provided below, list your major divisions in order of importance. Then, assuming you have 100 percent of your time to devote to them, give each area its ideal percentage—the amount of time you'd ideally like to devote to each area. Then in the last column, indicate what percentage of your time you actually spend on an average in each division. Consider starting with the three major divisions I consider mandatory—love and affection, self-development, and financial self-management. Then add four more of your own.

When you do this, feel free to read the list of sample major divisions chosen by other people as a guide, which I mentioned above. If you're coming up with more than seven major divisions, try grouping them. For example, if your list includes diet, running, and weight training, consider grouping them into a major division called "health and fitness."

Don't worry about selecting the wrong divisions. What's important for someone else may not be important to you. For example, I modified my own divisions from seven when I started to four. I changed the rank of my divisions once, moving a combined category of legal and finance from seventh to fifth when I realized I wasn't paying enough attention to it. Another time, I combined two previously separate divisions—family and friends—into one when I realized I considered all of them to be loved ones. Still another time I eliminated a division called "projects" when I realized that each one of these could fit into other divisions. Be creative and allow yourself flexibility as you process and progress over time.

You will find that you will end up with the divisions that increasingly mean the most to you.

Now contemplate your seven major divisions and write them below in order of most to least important to you. After you've done that, assign the percentage of time that you would ideally like to spend in each division. Then, in the next column, assign the percentage of time you actually spend in each division.

	Division	% Rank (Ideal)	% Rank (Real)
1.			
2.			
3.			
4.			
5.			
6.			
7.			

Finally, compare your ideal with your real life, noting in the column provided the percentage difference. For example, if for family and friends your ideal percentage was 40 and your actual percentage 10, the difference is 30.

	Division	Ideal	Real	Difference
1.				
2.				
3.				
4.				
5.				
6.				
7.				

You may notice quite a bit of difference between the time you'd like to spend on some things and the amount of time you actually spend on them. Chances are, those areas with the greatest discrepancies between real time and desired time are the areas where you're having the most difficulties. Don't be discouraged if you're way off; your brain is a self-correcting organ and will redirect itself if it knows when it's going off target. Remember that the brain works better at a conscious than at a subconscious level.

The purpose of this exercise is to examine the difference between your real and your ideal life, and to allow you to change the pattern so that your reality more closely approximates your ideal—the life you envision for yourself.

Exercise 2: Organizing Your Major Divisions

Now for the final step of this self-management process—the detailed organization of your major divisions. Although this requires some work, it will pay enormous dividends. With this step, you're going to organize your major divisions into seven sections—seven being the magic number for what people can effectively handle. Each of these sections is divided as follows:

1. IMAGE PAGE

For each of your major divisions, it is important to have an image that you have either drawn or selected from your own photography collection or from a book or magazine—an illustration that epitomizes for you the subject matter of that particular area. For example, for the self-development section, you might have an image representing quality time with family or exercising.

The function of this image is to trigger the imaginative right side of your brain and to give you a mental set that allows your self-correcting brain to aim constantly at an ideal. This is a technique used by many great and successful people including Albert Einstein, John D. Rockefeller, Henry Ford, Muhammad Ali, Björn Borg, Theodore Roosevelt, Wilbur Wright, Woodrow Wilson, Alexander Graham Bell, and Gary Player. The image gives you pleasure whenever you see it, which draws your mind's associations more positively toward that which you wish to accomplish.

It is advisable to draw your own image, because this gets the networks of your mind more engaged. As time progresses, your mind will encourage you to improve the image, at the same time improving your skill in art—a skill that feeds back to the right side of the brain, producing an overall development.

At your leisure, turn to the image page of Major Division 1 and copy it for the balance of your divisions, completing your ideal image for each one. Copy the pages for the remaining divisions as well.

2. MAJOR SUBDIVISIONS

When you have completed your image for each Major Division, you will be ready to divide each division into seven or fewer subdivisions. For example, the self-development subdivision could include vocabulary development, books, theater, quotes to remember, musical instruments to play, and conversational or presentation skills you'd like to develop.

At this stage, some people protest that there is no way they can do this, because they have far too much to accomplish in each division. If this is how you feel, it may well mean your life is running away from you. The fact is that there simply isn't the time in the average life to cope with more than seven subdivisions of seven Major Divisions; you must choose what you want to do *most*.

3. MAJOR GOALS

Next complete, either in image form or in order of priority, the major goals you wish to accomplish in each of your Major Divisions. Beside each major goal, put a rough estimate of the date on which you would like to have accomplished that goal. (I have yet to meet anyone who has been 100 percent successful in this estimation; as with the percentage rankings, it is a question of starting off on the path toward the ideal and allowing your brain to direct you, with its trial-and-error mechanism, toward the goal.)

4, 5, 6, 7. SPECIFIC GOALS AND DATES

In each Major Division, the fourth through seventh pages are for listing in chronological order the particular goals you wish to accomplish. This will enable you to keep a year-long check on your progress.

Failure to meet all these immediate goals is in no way a disaster; it is, in fact, common. When goals are not met, reassessment and reevaluation are required, and a more realistic date and/or realistic goal can be set. In some instances, time pressure and changes in your life may actually mean the useful discarding of certain goals.

While you complete each of these goal pages, bear in mind the fact that this is your life and that you arrange it as you wish. Your picture of your ideal life will be different from everyone else's. If, for example, certain aspects of your behavior reduce your life expectancy and you accept that knowingly, that is your own personal choice and not something you either "must" or "must not" do.

Similarly, be aware that your first subdividing, first ranking, and first goal setting will be just that—your first. As time progresses and the nature and tenor of your life change and develop, so will the importance you place on individual divisions. Certain divisions will become less significant or perhaps even disappear; other divisions that do not exist at the moment may suddenly enter and even come to dominate your life.

Major Division 1
Image Page

Copy this page to use for the next six divisions

Major Division 1
Major Subdivisions

Copy this page to use for the next six divisions

Major Division 1: Major Goals

Copy this page to use for the next six divisions

Major Division 1: Specific Goals and Dates

Goal	Date to Accomplish

Copy this page to use for the next six divisions

Major Division 1:
Specific Goals and Dates

Goal	Date to Accomplish

Copy this page to use for the next six divisions

Financial Self-Management

One of the major divisions of everyone's life is financial self-management. Regardless of how wealthy you are, you need to manage your income and expenses to ensure a positive cash flow—to ensure that you have enough income flowing in to cover your expenses flowing out. You also want a positive net worth: you want the value of what you own to exceed the value of what you owe. A positive net worth provides you with a reserve from which you can draw in the event that your expenses unexpectedly exceed your income.

As part of your own self-development, you must incorporate a financial self-management system that includes net worth and cash flow analyses. You need to know the dollar value of what you own versus what you owe (net worth), along with the amount of money you're earning versus the amount you're spending on a monthly basis (cash flow).

Amazingly, a recent survey discovered that 95 percent of all people avoid taking a hard look at their own personal financial resources and the flow of those resources. When they do look at them, however, they eliminate masses of stress, and they find other uses for the energy they have been spending on concerns about money. It really is crucial to get a good grip on your finances and apply your brain to your personal financial management in effective ways.

EXERCISE 3: NET WORTH ANALYSIS

Your net worth reflects your financial success. It shows how much money you would have if you sold everything you own and paid off all your debts. Think of it as your grade in the subject area of financial management. Your net worth should increase steadily over the earlier years of your life as you build wealth. It often decreases after retirement as your income from employment declines and you live more off your savings—spending the fruits of your labor to more fully enjoy your golden years.

However, your net worth need not decline during your later years. Many people continue to earn a respectable income after they retire from employment, businesses they own, or their investments.

Net worth becomes very important whenever you apply for a loan, such as a mortgage on a house. Prospective lenders will want to see that you have enough net worth (collateral) to cover the loan amount if your income is interrupted for some reason. Net worth also serves as a reservoir from which you can draw money in the event of an unexpected loss of income or unanticipated expenses.

What I *Possess*: To calculate your net worth, start by adding up the dollar value of all your savings and investment accounts and anything of value you own, such as your home, car, boat, jewelry, art, furniture, tools, and so on. The question you need to answer is this: if you sold everything you own, how much money would you have?

What I *Owe*: Next, list and total the amount of all your debts, including the balance you owe on your home, car, credit cards, and any other loans you have taken out.

What I *Possess*

Possession	Value
Total	

What I *Owe*

Debts	Amount
Total	

Knowing the value of what you own and how much you owe, calculating your net worth is easy. Just plug the numbers into the following formula and do the math:

Assets – Liabilities = Net Worth

EXERCISE 4: CASH FLOW ANALYSIS

Cash flow analysis is used routinely in business to ensure that the business is profitable—that it has more money flowing in from sales than flowing out in expenses. Managing cash flow is also important for you as an individual to ensure that you don't run out of money and don't rack up more debt than you can realistically make the monthly payments on.

Calculating cash flow is easy. The formula is very basic:

Income – Expenses = Cash Flow

To conduct a cash flow analysis, fill out the Cash Flow Forecast chart on the next page. Write all your monthly income in the spaces provided in the top half of the chart, under the month in which you expect to receive it. If you have a regular job and know that you're going to receive, say, $5,000 each month, your task is fairly easy. If you have seasonal income, you can confine your extra income entries to those months where it seems certain you'll get work. If you receive cash gifts for birthdays or anniversaries, enter those amounts into the proper month. Remember to enter the income in the month you expect to receive the cash. In the final column on the right, the total column, include any bank balances, CDs, or other investments. Total every vertical column.

Repeat the process for your expenses: monthly gas and electric bills, rent or mortgage payments, car payments, gro-

Cash Flow Forecast

For the Period from ________ to ________

Month	Jan	Feb	Mar	Apr	May	Jun	Jul	Aug	Sep	Oct	Nov	Dec	Total for Whole Period
Income Sources													
Total income for period													
Opening bank balance (if in credit) A													
Total B													
Expenditure													
Total expenditure for period													
Opening bank balance (if in debt) C													
Total D													
Difference between B and D													

ceries, fuel, travel costs, vacations, and everything else you spend money on.

After you total your expense columns, subtract total expenses from total income in each column and you'll see what you have at the end of each month. If the remaining amount is positive, you have a surplus, a positive cash flow. If the remainder is negative, you have a negative cash flow (more flowing out than flowing in), which is usually a sign that you need to tighten your belt.

Cash flow analysis helps you make key decisions on how to spend money, when to spend it, and when and where to cut back. Your cash flow forecast also enables you to identify periods when you can spend *more* to achieve your various life goals, including enjoying yourself.

Practical Uses for Cash Flow Analysis

Developing a cash flow forecast really can improve your life. For example, a friend of mine lived on a regular monthly income of about $4,000, which he considered a bare minimum for existence. Yet he recently had the opportunity to take off for a year and travel through Europe with a friend, all on a budget of $14,000. Now this friend had never been a particularly good money manager, and he avoided financial planning for his entire life, but circumstances forced him into it at this time, and he effectively became the financial controller for the two of them. He broke the money into weekly and monthly amounts, determined their essential expenses, and maintained tight control over the purse strings. A bank guard couldn't have done a better job of protecting their money.

To my friend's surprise, living on a strict budget was easy. In fact, he managed to stay below budget for a number of the weeks. When they were under budget, they rewarded themselves with special meals in fancy restaurants. When they noticed that they were spending too much, they froze their spending, staying put for a time and living off their supplies. They found that many of their best experiences, interestingly, were a result of staying right where they were. It was a superb exercise in financial and self-management that had some surprising benefits.

Unexpected events often bring unanticipated joy. Another friend of mine was on a business trip to Denver. She arrived on a Sunday to prepare ahead for two heavy workdays. A sudden snowstorm hit; she thought the snow would never stop. Roads were impassable. Telephone lines were down. She was stuck at her hotel with no way to reach home, her clients, her office, or anyone else she knew.

To make a long story short, she never had so much fun in her life. The hotel decided to make the best of a bad situation and keep the bar open for the next 48 hours. They provided free meals. The band that was stuck at the hotel seemed to play nonstop, and everyone got to know everyone else. The hotel guests and staff formed teams and had snowball fights, indicating that everyone involved was able to achieve balance.

Serendipitous events like these rarely happen to people who let their left-brain skills totally dominate their lives. Such people tend to overschedule themselves to the point of allowing life to become dull and monotonous, the opposite of what it should be. If you happen to be the kind who needs to schedule tightly, then schedule free time with nothing planned, as we did for our organization with the buffer zone concept. You must have free brain ranging time to allow your whole brain to function to your benefit.

This discussion itself is an example of a free brain ranging exercise. We were discussing cash flow analysis and somehow wound up talking about traveling through Europe on a shoe-string budget, enjoying an open bar, and engaging with fellow travelers in snowball fights. Very serendipitous.

In fact, it is nice to take a side trip as long as you don't lose track of your focus. At this point, our focus is on financial self-management. When you have your finances under control, you may find that you have surplus funds. When you know that these funds are available, you can plan your buying or investments or simply choose to spend your serendipitous windfall profit.

Investment Fundamentals

Financial experts have developed entire courses and written thousands of books on investing, covering everything from the basics to advanced strategies and techniques. I cannot possibly do the topic justice in the context of this book, but I do recommend that you allocate your investment dollars across the following three divisions based on relative risk:

- Savings in a bank account or other low-risk account that you can access quickly to cover unexpected expenses and emergencies. Think of it as a reserve fund.
- Lower-risk investments, such as bonds.
- Higher-risk investments, such as stocks—money you can afford to risk in pursuit of a significant return.

Putting technology to work for you

Managing your personal finances has never been easier. Personal finance software such as Quicken and digital transactions have made it easy to track income and expenses in every category down to the penny. Assuming you use a credit card to pay most of your bills, you can have your transactions automatically downloaded to your personal finance program, so you don't even need to enter transactions manually, although you may need to assign transactions to specific categories, such as groceries, gas, and dining out.

Most personal finance programs can use your transaction data to generate a broad range of reports for you, as well, including net worth, cash flow, budget reports, and many more.

In addition, credit reporting agencies, such as Experian, can provide you with your credit report and credit score, so that you can work on improving your credit ranking, which impacts your ability to borrow money, the interest rate at which you can borrow money, and possibly even your insurance premiums. Several smartphone apps are also available for managing personal finances and monitoring and improving one's credit score.

Dealing Effectively with Bad Financial News

Focusing some thought on finances seems prudent on its surface, but what if you're in a negative financial situation? If you discover that the bank is foreclosing on your mortgage, repossessing your car, or suing for your assets, that's likely to increase your stress level and degrade your overall sense of well-being. In other words, a focus on finances won't be good for your brain, right?

The solution is to deal with the situation. Financial self-management is one of the divisions of many people's lives that generate the most fear and stress. Even bright, healthy, interesting people get into trouble with their finances for a very simple reason: they're reluctant to look at the facts. Or they merely *hope* that things will sort themselves out financially; whenever the bills pile up, they just try to tune them out.

If you have ever been in financial straits, you know that ignoring the problem never helps. The problems persist, while the stress deepens. You wonder whether you'll ever be able to catch up on your bills, you're upset with yourself for getting into the situation, and you live in fear of a disaster occurring that will require a large cash outlay.

Incidentally, these fears grip people at every level. It's not uncommon to find individuals with respectable incomes and high net worth experiencing a sudden crisis due to poor financial self-management. It happened to a friend of mine, the president of a large company. One morning, my friend received a frantic call from his son at college. The registrar had told the son that the dad hadn't paid the tuition and other fees. The son wasn't allowed to attend classes until the registrar received the check.

My friend was able to gather the cash and make the payment, but having to suddenly come up with a few thousand dollars to pay an unexpected bill was a financial blow, and it was very embarrassing. Even though he was the president of a company, he simply hadn't done any personal budget planning. You can imagine how his stress level must have soared as he worried about what might happen again in the future, because this was a really unanticipated financial blow.

Exercise 5: Time Management

Managing your life isn't all about money management. It also involves the effective and efficient management of time. Most of us waste a considerable amount of time engaged in unproductive and ultimately unsatisfying activities. We watch too much television, spend too much time on social media, and are glued to our smartphones, entirely neglecting the wonderful real world that surrounds us and depriving ourselves of real-life experiences and adventures.

To see how much time you're wasting, devote the next week to logging how you spend your time. Keep track of how much time you spend on work, self-development, and other productive, meaningful activities and how much of your time you waste on watching TV, texting, playing video games, and so on. Start with the chart on the following page, and modify it to reflect what you spend your time on.

The purpose of this exercise is to see whether you can free up time to spend on more productive and rewarding pursuits. If you're logging a lot of time on leisure and recreational activities but very little time with family and friends, you may have an imbalance in your life that needs to be addressed.

Activity	Monday	Tuesday	Wednesday	Thursday	Friday	Saturday	Sunday
Work							
Sleep							
Personal Care							
Housekeeping							
Study/homework							
Volunteer							
Exercise/recreation							
Meal prep/meals							
Leisure							
Reading							
Shopping							
Commuting							
Family/friends							

CHAPTER 7

Learning How to Learn

In a time of drastic change,
it is the learners who inherit the future.
—ERIC HOFFER

We all assume that learning is natural. After all, we have an innate desire to learn how to crawl, walk, and communicate, and we learn each of those complex skills without any formal education or training. Yet when the time comes to learn math, spelling, English grammar, science, history, or a host of other subjects and skills, we often struggle. We even have difficulty in our careers and businesses keeping pace with the persistent flood of information. But we are living in the information age, so being able to learn and process information has become nothing less than a survival skill.

Given the learning challenges we face to remain competitive in the business world, to function well in our personal lives, and to fully enjoy all that life has to offer, we need to develop better ways to learn.

In this chapter, I describe some of the common obstacles to learning, offer guidance on overcoming these obstacles, and present several exercises to help you optimize your innate ability to learn.

The Reluctant Learner

The biggest obstacle to learning is reluctance. To illustrate my point, I share my story of the reluctant learner. He can be described as a student who has a major exam coming up, or a businessperson who needs to review materials for an important meeting or report that's due shortly. See if this resembles someone you know.

At about six o'clock in the evening, our best-intentioned individual, with whom you may be all too familiar, sits down at his desk and gets everything ready for the main event that is to follow. He gets his papers in order, lines up pens, pencils, and erasers, puts the radio on low in the background, and adjusts the books—everything in perfect order.

Now that everything is its proper place, he does it all again. Little adjustments here and there, sorting through his papers, lining up the writing tools, shifting the desk lamp to a slightly better position, and so on. During this delay process, he remembers that he didn't quite finish reading this morning's newspaper. As he reads the news, he suddenly finds a certain article much more interesting than when the newspaper arrived that morning. He thinks, "I'd better finish reading this one article to free my mind so I'll be able to concentrate."

As he nears the end of the article, he can't help noticing a few other interesting items he had missed. And will you look at this on the entertainment page? There's a television show, and he needs to watch it, because if you're going to work during the evening, you need some kind of break. The show starts at 7:00, about 15 minutes from now, so there's no sense starting on the book. Besides, he could use the rest and relaxation before he really gets cracking. He figures he'll watch the first 15 minutes of the show and then get to work. By 7:15, he's fully engrossed in the show, so he continues watching it to the very end, at 8:00.

As our Mr. Best of Intentions returns to the desk, he does an amazing thing. He cracks open the book, but then remembers a phone call he was supposed to make arrangements with Bill to go to the basketball game on Friday night. If he doesn't call right away, they might lose out on the tickets. As luck would have it, Bill had a lot more to talk about than just the basketball game. Like the newspaper and the TV show, the conversation is far more interesting than anticipated.

Finally, at 8:30, our Mr. Best of Intentions readies himself at the desk and sits down to the book, and this time he means it. He opens the book to page one, thumbs it, starts to read it, and realizes, guess what, he's just a little bit hungry and a little bit thirsty. He knows from experience that if he doesn't take care of that right away, he'll never be able to concentrate.

Off he goes to the kitchen, he opens the fridge, and there it all is. The first temptation leads to another, which leads to another. Before you know it, what started out as a light snack becomes a massive meal, but at least he's got that out of his system. Now he returns to his desk, absolutely certain that he's ready to tackle the important task that faces him, and it's about 10:30. Feeling a little bit guilty, he decides to go on until maybe 12:00 or 12:30 the next morning.

Once again, he opens to page one and rereads the first sentence, but he realizes that all that energy that should be up in his head is down there in his stomach. He's feeling a little bit tired. Why not watch the 10:30 show that he likes so much and give himself time to allow that food to digest? Then he'll really be able to get down to the work at hand. At midnight, there we find him, sound asleep in front of the television.

How will Mr. Best of Intentions react when someone wakes him up and sends him off to bed? He'll probably think that he really cleared the decks, even though he didn't get his reading done. He relaxed, rested, filled his belly, was entertained, and took care of the plans for the basketball game. Oh well, tomorrow at 6:00 p.m., he'll really get down to work.

Sound familiar? We've all been there. We've all done that. But there's a better and more efficient way to approach any study project.

Fear of Learning

The story of the reluctant learner is certainly familiar to all of us. On one level, it is amusing. On another level, it demonstrates how creative we can be in coming up with good reasons to avoid what we find to be unpleasant. We can come up with all sorts of excuses for not putting forth the effort to learn something, such as the classic, "My dog ate my homework."

We're very inventive when it comes to finding alibis, which is proof that we have the ability to be creative in other areas as well. But familiarity with this situation is also discouraging because it illustrates the fear that so many of us associate with any sort of academic study. I use the word "fear" intentionally. The problem isn't restricted to not liking to read textbooks and reports. It's fear.

The fear goes back to when we were given textbooks in school: we knew they were harder than the story books we loved to read. We knew they meant work, and we knew we were going to be tested on the information in the book. We never totally get over that fear. In fact, the threat of testing can totally disrupt the brain's ability to work under certain conditions. Countless people with "test anxiety" freeze during tests, even though they know the material perfectly well. I know a story of a student who had a panic attack during one test, which was totally debilitating: she couldn't remember a thing. The minute she left the room, it all came back to her.

There are documented cases of bright people who spent an entire two-hour test period furiously writing, believing they were answering the question, while actually they were writing their own name or a single word over and over again. Now that's fear, or perhaps better, terror. They know that if they study and perform poorly in school, they will feel ashamed and embarrassed. They will feel that everyone, including their teachers, fellow students, and parents will think they are stupid. The child may actually begin to believe that he's not bright enough to handle the material.

But children are clever. They figure out ways to avoid the psychological and emotional pain of performing poorly at school. In fact, what many students learn in school is how *not* to study in order to avoid the potential fallout from failure. If he doesn't study and doesn't do well, so what? Who was interested in that stupid stuff anyway? Avoiding the work presents no threat to his self-esteem. He has a reason to explain his failure, and it is reinforced by his peers. He's showing how brave he is in a situation that is terrifying to them. He becomes a hero. In other words, a child who avoids work is lionized because he stands up to a fear every other kid has. That's why those kids are often looked up to as leaders. They're rewarded for a kind of display of courage.

Even people who force themselves to study will hold on to a part of themselves that emulates the behavior of the ones who don't. Kids who get 80s and 90s in their tests often use the same excuses for not getting 100s that the nonstudiers make for failing.

Secrets of the Super Learners

In traditional education, the information has been given to the child based on the assumption that information should flow *from* the subject *to* the individual. Information is, in a sense, thrown at the student, who is expected to absorb, learn, and remember all of it. Recent discoveries about the brain have shown that this is not how we learn best. We place more emphasis on the information than on the individual, which is wrong and ineffective.

Everyone in today's society is swamped with information; the volume of it is unimaginable. We're asking people to cope with this information explosion with the same outdated skills they've been using for centuries. We've got to change. If we're ever going to learn to cope with the knowledge and information that we need—a data flow that will increase dramatically as we progress—we must take advantage of what we now know is inside all of our brains.

We must reverse our previous process. We must take advantage of our natural ability to learn, think, remember, recall, create, make decisions, and solve problems. We need to place the emphasis on *us* and then allow *us* to get the information in the appropriate brain-relevant way.

That's where the major change in information management breaks through. Instead of first learning facts about other things, superachievers begin by learning *how* to think, *how* to recall, *how* to create, *how* to solve problems. The superachiever is the one who has *learned how to learn.* Alvin Toffler, the author best known for *Future Shock,* said in his book *Power Shift* that in the future, the illiterate person will not be the person who cannot read. It will be the person who does not know how to learn in order to learn.

The traditional school approach is a grid technique that comprises a fixed series of steps that must be followed, regardless of the subject matter. Some teachers actually suggest that a difficult text be read three times to ensure a total understanding, but they never provide a focus for the reading. It may make sense to read something three times, but only if you know how to shift gears and read in different ways for different purposes.

You can't learn everything the same way. There's a world of difference between studying a text on literary criticism and a text on advanced calculus. The key is to start by working from the individual outwards. In other words, if I wanted to teach you advanced calculus, I wouldn't start by throwing books, formulas, theories, and texts at you. I would start by teaching you how to learn the material—by teaching you how *you* learn. You would need to start the learning process from within. You have to know how your eyes work when you read. You have to be able to retain and recall information. You have to learn organizational skills, problem-solving techniques, and how to use Mind Maps.

Everything these days comes with an instruction manual, but we've never been given an instruction manual for ourselves. Think about it: How does your brain work best? What can you do to make yourself more efficient? That's what this program does for you. It provides a maintenance and instruction manual for the most complex organ on earth—your brain.

The great Russian researcher Dr. Pyotr Anokhin studied the information processing capabilities of the average brain. He described the pattern-making capability of the brain as "so great that writing it out would take a line of figures in normal manuscript characters more than 10.5 million kilometers in length. With such a number of possibilities, the brain is a keyboard on which hundreds of millions of different melodies, acts of behavior, or expressions of intelligence can be played. No person yet exists, or has existed, who has ever even approached using the full brain. We accept no limitations on the power of the brain. It is limitless."*

Everything in my brain power system is designed to remove the boredom from learning and allow that supercomputer we call the brain to do the work it was designed to. Which takes us to the seven steps to becoming a super information processor:

---

* Anokhin was correct. The brain is, we now know, scientifically limitless.

1. Browse. Get an overall feel for the material and sense if the material is heavy or light.

2. Allocate study time and quantity. Time and work limits increase your comfort and confidence levels.

3. Mind Map the topic you're about to study. A Mind Map helps you assess your current knowledge about a topic or what you think you know about it, which makes you more receptive to the information.

4. Ask questions and define goals. Describe exactly what you want to learn from the material.

5. Skim the material. Read the first and last paragraph of each chapter, look at the charts and illustrations, and skim the glossary for any unfamiliar words.

6. Conduct an inview. Read what's most relevant to you and what you think is most important, skipping anything you already know or that's irrelevant or unimportant.

7. Review. Reread anything you're unclear about or that you overlooked. Do you have any unanswered questions from step 4? Have you met your clearly defined study goals?

The following sections explain each of these steps in greater detail.

STEP 1: BROWSE

Learning begins best with preparation, which can be compared to the process of selecting a book to read. You may look at the front cover, read the back cover, and skim the book jacket to get a feel for the story line. Maybe you read quotes from critics and reader reviews, or you thumb through the pages.

You're trying to decide whether the book is something you want to invest your money and time in.

Is that the way you approach a book you want to use for a research project—a management book, for example, which contains numerous theories that deals with the hierarchy of needs and other complex human relations topics? Suppose your boss hands you a book and tells you to read and summarize it. You probably just dive right in, reading it from cover to cover.

With a book on a complex topic, taking the time to get a general idea of what the book is all about is even more important. It can make the reading simpler, faster, and much less boring. In fact, it can make the reading fun. If you think that claim is a stretch, I'll let you judge for yourself as you put into practice the techniques I'm about to teach you. I'm sure that you will discover what thousands of others have experienced: this overall system will take the monotony out of your reading and will remove a major block that stops you from doing what you're required to do.

Whether it's a book, a research report, a white paper, or whatever, spend some time getting a feel for it. Is it difficult or easy? Do you have any knowledge of the subject matter? Are there diagrams and charts that will help you as you go through? How much, in relation to the text, do you need to know? Are there summaries, is there an index? Become familiar with the book without trying to read it word for word.

As you become familiar with a book before reading it, your brain begins to build the underlying framework for organizing the information presented. The table of contents, for example, provides an outline for the book. It's the author's way of saying, "Here's what I think is important." If the book has an index highlighting key topics, it too reflects what the author or indexer believes is most important. In many ways,

A success story

A student at Oxford had spent four months struggling with an exceptionally difficult psychology text. With fifty pages to go, he told me he was "starting to lose it" because the amount of information was bogging him down. As he put it, he was "drowning with land in sight."

As it turned out, he was having that problem because he was reading in the way we've all been taught to read: start at page one, and slog your way through. Although he was just about at the end of the book, he had no idea what its last chapter was about. And what do you think it was? It was a total summary of the book, with all the key points, all the major graphs and formulas: everything he really needed to know.

After reading that chapter, this fellow estimated that he could have saved himself at least seventy hours of reading time, perhaps twenty hours of note taking, and months of worry and stress by simply knowing how to use his brain. All he needed to do was follow step 1 of my five steps to becoming a super information processor. Just imagine what he could accomplish by following all five steps!

the index is even more of a guide to the book than the table of contents, because the index highlights key words and concepts in greater detail.

STEP 2: ALLOCATE STUDY TIME AND QUANTITY

You can make an overall study assignment much less overwhelming by following the simple strategy of divide and conquer. Think back to the reluctant learner described earlier in this chapter. How do you think his story would have been different had he selected 12 pages to complete in 45 minutes? He would no longer be intimidated by the thought of getting through a 200-page book, which is a formidable feat under the best circumstances. Now he would be looking at getting through 12 pages in less than one hour, which would be much more manageable.

We can deal with chunks of time and work much more effectively than when we approach what appears to be a monumental project. Chunking involves sitting down to work and deciding how much time to devote to the project and how much content to cover in that time.

Why does this method work better than just sitting down and going at it? I'll answer that question after you complete the following exercise.

Exercise 1: Shape recognition

Write the name of the shape next to each number (see Figure 7-1).

Gestalt psychologists believe that our brain has a tendency to complete things. For example, you probably labeled the shapes in exercise number one as follows: Straight line,

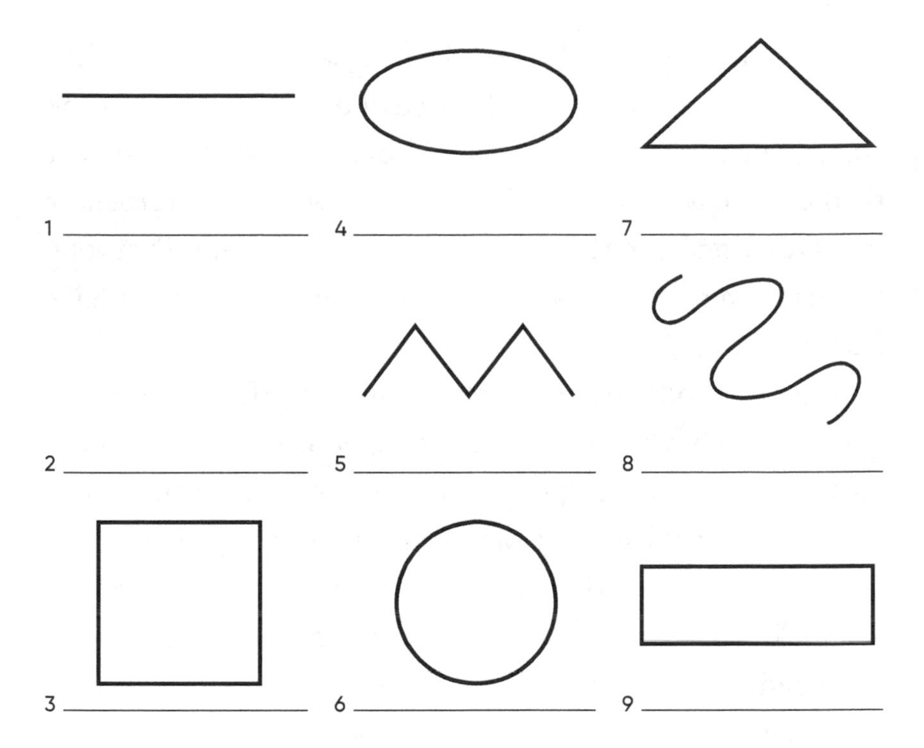

Figure 7-1: Shape recognition.

cylinder, square, ellipse or oval, zigzag line or some variation, circle, triangle, wavy or curved line, and rectangle. But did you notice anything about the circle?

Many people see the circle as a complete circle. Others see it as a broken circle, but assume it was supposed to be complete. We like things to be neat and tied up.

When we set targets or goals such as so many pages in so many minutes, we see a point of completion. We have something to shoot for. There's a link between start and finish that encourages us to complete the job rather than wandering off somewhere.

An audience responds to a lecture or presentation best when you take the following approach: tell them what you're going to say, say it, and then tell them what you said. It also helps to deliver the talk in a limited and predetermined time frame, such as 45 minutes, and communicating that fact to the audience beforehand. In these ways, the presenter prepares them for the learning that is to come. Everyone processes new information better when guidelines and a framework are in place.

One method I use is to physically set off the pages I'm going to study with a big sheet of paper at the front and the back of the section I'm going to tackle. Creating a limited section of pages provides two practical benefits:

- Navigating a section is easier than navigating the entire book. You can flip back and forth in a section without spending much time trying to relocate where you were.
- You eliminate the fear of the unknown. There's a specific amount of work to be done. You can see it and feel it. An entire book is threatening. Studying in chunks is not. There's an attitudinal difference at play here that makes a performance difference.

Now suppose you've blocked out a two-hour study period, and at the end of the second hour, you're on a roll—everything's starting to click. What would you do at this point? Keep going now that everything's starting to mesh or take a break? Remember the activity-rest cycle introduced in chapter 1. Even if you're making good progress, it's important to take a rest to give your brain time to process the information and integrate it.

STEP 3: MIND MAP THE TOPIC YOU'RE ABOUT TO STUDY

For the brain to process and integrate information most effectively, the information must be structured in such a way as to slot it in as easily as possible. If the brain works primarily with key concepts in an interlinked and integrated manner, our notes and our word relations should be structured in this way rather than in a traditional linear fashion.

Rather than starting from the top and working down in sentences or lists, notes should start from the center, with the main idea, and branch out as dictated by the individual ideas and general form of the central theme, as shown in the image below (Figure 7-2), which is a Mind Map for the topic of space exploration.

A Mind Map is simply a tool for engaging all your cortical skills. You use logic, words, lines, and lists, as you've always done. Then you add the right-cortical skills of image, color, rhythm, and dimension. Now on the page in front of you, instead of the standard boring linear gray and black list, is a colorful, graphical, multidimensional map of your thoughts.

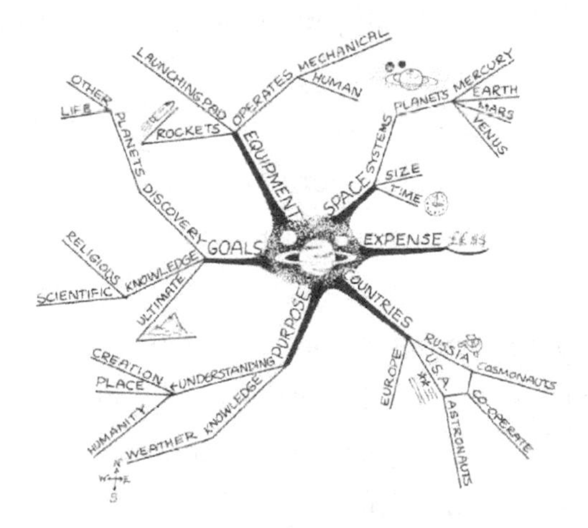

7-2: Initial ideas Mind Mapped around a central theme.

A Mind Map has a number of advantages over linear note taking.

1. It uses color, which improves memory and recall. Although the figure above is in gray scale, I strongly encourage the use of different colors, including a colorful central image.
2. It uses images, which engage the right brain to enhance memory and recall, including a central image for the main idea.
3. The relative importance of each idea is clearly indicated. More important ideas will be nearer the center, and less important ideas will be near the edge.
4. The links between the key concepts will be immediately recognizable because of their proximity and connection.
5. As a result of the above, recall and review will be both more effective and more rapid.
6. The nature of the structure allows for the easy addition of new information without messy scratching out or squeezing in.
7. Each Mind Map is unique, which enhances memory and recall.
8. In more creative areas of note making, such as essay preparations, the open-ended nature of the map enables the brain to make new connections far more readily.

Mind Mapping engages your entire brain to help you continue to learn as you study, and it is the key to becoming a super information manager and processor.

I discuss Mind Mapping in greater depth in chapter 9. For now, you shouldn't spend more than two to five minutes on your knowledge Mind Map. As you're preparing for a study session or reading a book, you are simply using a Mind Map to improve your concentration and create a positive mindset, filling your mind with important information relevant to the topic and not wasting time on extraneous material.

In this way, you're forcing yourself to explore everything you know about the topic before you actually read about it. This two- to five-minute step prepares you for the learning that's to follow. It gets you thinking about the topic at hand instead of the movie you're going to see later that evening or the friend you want to phone. It keeps you focused.

If you know nothing about a topic, Mind Map whatever you *think* you know. Whether you're right or wrong doesn't matter. What matters is that you're getting the engine cranked up for the main event. You're focusing your brain's power of association on the task at hand.

Another benefit is that as you make this practice part of your study routine, you're going to develop better recall overall, because you're taking regular inventory of what your brain knows. You're going to be able to access your data instantaneously from your entire vast range of knowledge. You will surprise yourself at meetings, conferences, school, and every situation you're in with the brilliance of your ability to retrieve what you've stored in that amazing brain of yours.

STEP 4: ASK QUESTIONS AND DEFINE GOALS

This fourth step may strike you as odd. After all, how can you be expected to ask questions and define your learning goals if you haven't even read the material?

Let me answer that question with a question: how does a scientist prove or disprove a hypothesis? She questions it: "What will happen if I move this to here? Why should there be a reaction? Where will this reaction occur? Who can benefit from this finding? How can I convince the medical community that this process can be replicated? When are the conditions perfect for the tests?"

Take a look at the Mind Map presented in the previous section. You can write your question words right on the map. For example, on the bottom right branch, next to "Europe," you might write, "Who?" to indicate that you want to know specifically which countries in Europe are participating in the space program. In the branch marked "Cooperate" (between "Astronauts" and "Cosmonauts"), you might write "How overcome language?" Don't waste time or space with insignificant words, such as "a," "an," or "to," which merely clutter the map (and your mind).

If, as you're reading, you discover other avenues you want to question, follow them. Learning is a living process. As you read or study, you'll be refining your questions and learning goals anyway, but the more accurately you learn how to use questions—who? what? where? when? why? how?—the better you'll be able to perform.

Most textbooks used in schools have it all wrong; they put the questions at the end of the chapter, when they should be placing them at the beginning. However, some textbook publishers have the right idea. They're realizing that learning is enhanced by giving the reader a clearly defined purpose. The old days of just plunging in on page 1 are starting to give way to new discoveries about how we learn best. We too need to change the way we process information, at least to some degree.

EXERCISE 2: WRITING INSTRUCTIONS

One of the best ways to become more sensitive to how you learn is to try to teach someone something you know. This exercise gives you an opportunity to do just that. Imagine that you're an avid jigsaw puzzle fan. They give you great enjoyment. A good friend shows up one day with an enormous box. It's gift-wrapped and tied with string. She tells you that the gift has been called "the most beautiful and complex puzzle ever devised." You thank her for her thoughtfulness and decide you're going to go right at this wonderful new challenge. You're going to devote yourself to the completion of this puzzle now.

To complete this exercise, write step-by-step instructions on how to complete the puzzle, starting with the moment your friend walked away and left you at the door with the gift.

Steps to Completing the Jigsaw Puzzle:

1. ______________________
2. ______________________
3. ______________________
4. ______________________
5. ______________________
6. ______________________
7. ______________________
8. ______________________
9. ______________________
10. ______________________

11. ____
12. ____
13. ____
14. ____
15. ____
16. ____
17. ____
18. ____
19. ____
20. ____

Here's a sample from one of my students:

1. Go back inside the house.
2. Take the string off the box.
3. Unwrap the box.
4. Dispose of string and paper.
5. Look at the picture on the outside of the box.
6. Read the instructions, concentrating on number of pieces and overall dimensions of the puzzle.
7. Estimate and organize amount of time necessary for completion.
8. Plan breaks and meals!
9. Find surface of appropriate dimensions for puzzle.
10. Open box.
11. Empty contents of box onto surface or separate tray.
12. If pessimistic, check number of pieces!
13. Turn all pieces right side up.
14. Find edge and corner pieces.
15. Sort out color area.
16. Fit obvious pieces together.

17. Continue to fill in.
18. Leave "difficult" pieces to the end (because as the overall picture becomes clearer, and the number of pieces used increases, so does the probability that the difficult pieces will fit in much more easily when there is greater context).
19. Continue process until completion.
20. Celebrate!

Chances are good that your instructions are very similar to my student's instructions. The point here is that while you may begin on page 1 of a book every time, there's no reason to. Would you start a jigsaw puzzle only from the bottom left corner and force yourself to try to build it only from that point?

Probably not. Instead, you might start by looking at the picture on the box so that you have an overall goal in mind. You might organize the pieces into colors and patterns. Then you may place the four corner pieces and complete all the edges: top, bottom, left, and right. You can then work from the outside in or the inside out. And you're likely to reach for the easy pieces first—the low-hanging fruit. You can take a similar approach to reading a difficult book or report. Start with the overall picture, which becomes clearer and clearer as you fill in more "pieces."

STEP 5: SKIM THE MATERIAL

With step 5, you enter the application phase of super information processing. In this stage, you're trying to get an overview of the material. Using a book as an example, you want skim the parts of the book that are not part of the main body of reading, such as the following:

- Summaries and conclusions, which encapsulate what's covered in each chapter.
- Boxed text, which is often set off from the main text to highlight its importance.
- Glossaries, which introduce and define unfamiliar concepts and terminology.
- Tables, charts, photographs, and illustrations, which often present information in more accessible formats.
- Chapter and main section headings, which provide a mental map of how the information is structured.

These components of a book or report help you complete the central image of your Mind Map and to add the main branches that flow from that central image.

As you read, concentrate on the beginnings and ends of paragraphs, chapters, special sections, and even complete texts, because that's usually where the critical information is concentrated. A scientific journal, for example, may contain a lengthy article about an experiment. Quite often the first paragraph summarizes the study, and the last one provides an overview of the conclusions drawn from the study. If the results were not conclusive, you may not want to read the article at all, but at least you have a clue to whether the material will be of value to you.

Skimming is even more important before reading relatively brief reports or manuscripts that don't include footnotes, statistics, tables of contents, or summaries. Every piece of prose writing has paragraphs. Read the first sentence in each. Usually, the writer puts his or her key thoughts in what teachers used to call a *topic sentence.* You can look there for a sense of organization and a feeling of what the writer considers important.

The first and last paragraphs tend to set up the reading and summarize it. By keeping all these tools in mind, you can pull them out and use them as they apply. Being able to apply two or three of them is better than having no clue whatsoever and reading blindly.

One more suggestion: as you skim, use a pen, a pencil, a chopstick, a ruler, or some other object as a visual tracer to follow the words, charts, tables, and illustrations you're taking in. In the next exercise, I explain why.

EXERCISE 3: TRACING YOUR EYE MOVEMENT

Look at the graph below (Figure 7-3).

Figure 7-3: A simple graph.

You may think that your eyes have captured a perfect image of the graph—that the image is seared into your memory like a photograph, but what you are perceiving can be deceptive. Your eyes tend to fixate on certain areas and then move away. The image below (Figure 7-4) reflects the actual pattern of unguided eye movement. As you can see, the pattern of eye movement conflicts with the pattern of the line in the graph above. As a result, the image recorded in your mind is not as clear and distinct as you might think.

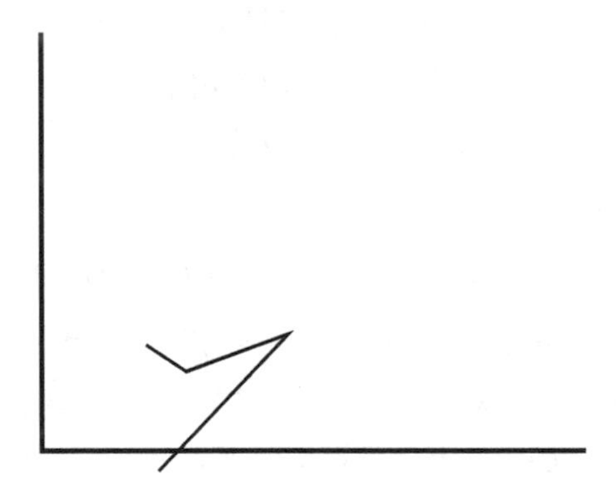

Figure 7-4: Standard pattern of unguided eye movement on graph, causing conflicting memory of shape of graph.

Next, using a pen, a pencil, a chopstick, or some other pointing device, trace the line on the graph. Now your eyes register not only the static impression of the line but also the movement and flow of the device you used to trace that line. You now have muscular memory, visual memory, and your standard memory all helping you remember what you have seen.

Using a tracer improves memory phenomenally. With the help of such a visual aid, eye movement more nearly approximates the flow of the graph, and the memory is strengthened by each of the following inputs:

1. The visual memory itself.
2. The remembered eye movement, approximating the graph shape.

3. The memory of the movement of the arm or hand in tracing the graph (kinesthetic memory).
4. The visual memory of the rhythm and movement of the tracer.

The overall recall resulting from this practice is far superior to that of a person who reads without any visual guide. It is interesting to note that accountants often use their pens to guide their eyes across and down columns and rows of figures. They do this naturally because any very rigid linear eye movement is difficult to maintain with the unaided eye. This is another example of how knowing how your brain and body function helps you to function better.

STEP 6: CONDUCT AN IN VIEW

With Steps 1–5, you've basically completed an overview and a preview of the material. These steps put you in control of the resource, so that you can work out in your own mind the best way to approach the material. That is, you can choose what to read and how you want to read it. Not everything in a book or report is relevant, just as everything a lecturer says, or what's on television in a demonstration or documentary, may not be relevant to you.

Take the same approach with a book as you do with a poor presenter or a TV program that's turning you off. Tune out the parts that don't matter to you. Focus on those that do. You may feel as though you're breaking an unwritten law by not reading everything, but you'll get over it.

With the *in view*, you're gathering information to fill in the gaps in your understanding. Think of it as the part of the process of assembling a jigsaw puzzle that you do after you've laid out the borders and established the color areas. You may have already gleaned everything you needed, and you just want to check out other information.

The value of the in view is largely when you're dealing with difficult material. You check out areas in which you may have some information gaps. As you know, if you give the brain a rest period and if you don't struggle with difficult material, the brain can often come up with an answer later—sort of a "how didn't I see that earlier?" response. Also, by having a better grasp of the surrounding material, which you've obtained through all the other steps, you can take advantage of the brain's tendency to fill in the gaps.

The in view can also make studying a more creative process. Just as Einstein jumped over a tremendous number of sequential steps by leaving out small amounts of material, you give your brain much more leeway to use its natural ability to create and comprehend.

STEP 7: REVIEW

Review is the final step. You've completed an overview, a preview, and an in view. If you need further information to answer questions or complete any of your goals, conduct a review. It's simply a matter of filling in anything that's still incomplete and rechecking the areas you considered important. Typically, the first six steps get you at least 70 percent of the way toward your study goal, leaving you with an excellent grasp of the material. Reviewing should involve no more, and usually far less, than what remains.

CHAPTER 8

Listening, Note Taking, and Speed-Reading

We now accept the fact that learning is a lifelong process of keeping abreast of change. And the most pressing task is to teach people how to learn.

—PETER DRUCKER

Most people will agree that the loss of one sense, such as sight, greatly sharpens that of another, such as listening. But current research indicates that this is not quite correct. It may be true that the loss of one sense forces us to develop the remaining senses to the maximum, but the more senses we have, the greater the potential ability of each. Likewise, the more techniques we have for receiving and organizing information, the greater our overall potential for learning and retaining that information.

In this chapter, I cover three more skills that will make you a super information processor: listening, note taking, and speed-reading. They're all incredibly valuable but are usually ignored in formal training.

Listening

Studies show that we spend 50 to 80 percent of our daily lives communicating. In the business world, listening is often named as one of the three most important managerial skills, yet it is the least taught of all communication skills. Hearing is so natural that we take listening for granted, but the two are not the same. Most people can hear, but they don't listen. They don't pay attention or make an effort to process and understand what they hear.

How would you grade yourself as a listener? Would you call yourself superior, excellent, above average, average, below average, poor, or terrible? Think about it for a moment. Give yourself a score between 0 and 100. How do you think your best friend would rate you? Higher? Lower? The same? How about your boss? Your business colleagues? Your subordinates on the job? How about your wife or husband?

I think you'll find the answers very interesting. Out of every 100 people who rate themselves as listeners, fewer than 5 percent rank themselves as superior or excellent. A whopping 85 percent consider themselves average or below average. On the zero to 100 scale, the average rating is a 55. People think their colleagues and subordinates would rank them about the same, and that their friends and their boss would rank them higher, possibly because we tend to pay more attention to our friends and authority figures. Early in a marriage, partners rank each other higher than they rank themselves, but as time goes on, the scores they give each other drop well below the scores they give themselves.

What's important is that we all have a great deal of room for improvement, and when we improve our ability to listen, we improve all our senses and our ability to process information at a much higher level.

Overcoming the Five Factors That Impair Listening

To sharpen your listening skills, first recognize the five factors that stand in the way our ability to listen and process the information we're hearing:

1. Physical impediments
2. Distractions
3. Boredom
4. Forgetfulness
5. Indistinct audio

In the following sections, I explain each of these five factors in greater detail.

1. PHYSICAL IMPEDIMENTS

Let's look at the physical problem first. Other than disease, hearing problems stem from our mistreatment of our incredibly delicate hearing apparatus. If we treat it like a very expensive, highly sophisticated musical instrument, it will remain in good order, so we should treat it as such. Here are two very important precautions to take to protect your hearing:

- Never stick anything smaller than your elbow in your ear to clean it.

- Avoid sources of loud sounds, including concerts, airplanes, and industrial processes. And don't crank up the audio on your stereo, earbuds, or TV. Talk and encourage others to talk in softer tones—no yelling or even raising your voice.

Studies show that avoiding loud sounds helps to preserve hearing. For example, the Mabaan tribe, near the Sudanese border, has a very gentle speech pattern, and they take a great care not to cause any shock to their hearing apparatus. They also have no decline in hearing ability with age; the hearing of their older members is as acute as that of their younger members. The more you can baby your hearing, the better it will remain.

2. DISTRACTIONS

The second main impediment to listening is distractions, which can be broken down into environmental (external) and mental (internal). Environmental distractions come from background sounds, such as traffic, crowds, machinery, or music. We have an amazing ability to filter out background sounds. Have you ever attended a noisy party where the hostess perks up her ears and says, "Excuse me a moment; the baby's crying"? Likewise, two lovers at the same party have no trouble carrying on a conversation in the midst of the hubbub.

We've all been in situations that are very noisy and chaotic in the background when the brain automatically screens out all the noises that threaten to interfere with what we're listening to. Just by becoming aware that your brain does this, you'll already begin to be able to eliminate many environmental distractions that otherwise would have interfered with the information you're most interested in processing.

The next time you're in a crowd, be aware of what you hear and what you don't. You'll find that if you focus your attention on what you want to hear, you may not even notice the background noise. You can guide your focus to a conversation that's important to you. Notice how your mind and body set themselves to listen, and then practice that ability whenever you're in a place where the noise level seems intolerable, or when you want to focus your attention on something especially important to you.

Unfortunately, our ability to tune out audio input also makes us more susceptible to selective hearing—hearing only what we want to hear. Some children are masters of selective hearing. They simply ignore whatever their parents, teachers, and other people in positions of authority tell them. However, people in long-term relationships also develop this ability, which can wreak havoc on their relationship. It's not that they don't hear one another; it's that they don't listen.

Selective hearing can be beneficial. Have you ever slept through incredibly loud noises, but then awakened immediately to someone's gentle, caressing voice? That's another example of the power of selective hearing, and exercises are available to help develop this skill. For example, if you're on a busy street, just stand there and listen for different sounds. First, try to pick out, say, the sounds of different shoes on the pavement. Then screen out everything else and listen for the sound of cars. Their tires. The unique sounds of their engines. Horns blaring. Then screen out all that, and listen only for birds, their songs, and even their flight. Can you tell a bus from a truck? Can you tell one brand of car from another?

You can do the same in your home. Can you hear the sound of electricity? Water? Wind? Movement in the house? Your pets? Other people? Try hard to distinguish different sounds. It'll be rewarding, and it'll be incredibly beneficial to you. Even silence has a distinct sound. Sound engineers can often listen to the silence during pauses on an audiotape and tell you which studio they were recorded in.

Distractions can also be internal. These often become a problem when you're tired or stressed. During these times, you may be carrying on internal dialogues with yourself or watching scenes play out in your mind that distract you from what's going on around you and what you're hearing. Sometimes you can become so engrossed in thought that you place the rest of your life on autopilot. You can drive to work or school or the grocery without giving a single second of conscious thought and arrive at your destination with no memory of how you got there.

One of the best ways to overcome internal distractions is to practice mindfulness: living in the moment. Stop thinking about the past and the future, and focus your mind on what you're doing at this instant in time. Therapists refer to it as "being present."

You can also eliminate internal distractions by using what you learned about training yourself to recall information—primacy, recency, linking, outstandingness, and review, which are all covered in chapter 4. Memory and listening skills are all linked in one way or another. If you put them to work, you'll eliminate self-made distractions.

3. BOREDOM

A common obstacle to listening is boredom. We perceive nothing interesting, relevant, or entertaining in what we're being exposed to, so we tune it out. If you've ever had to sit through a boring speech or presentation, you know the feeling: your mind tends to wander, and you don't listen to what's being said. Our lack of attention is a built-in defense mechanism against boredom. Unfortunately, we can't always just walk away, and we may need the information that we find boring, so we must have a solution to make what's boring seem less boring.

One cure for boredom is to challenge yourself to fully understand what's being said. As the person is speaking, don't just take notes; jot down questions as well. Make your notes entertaining by putting what the speaker says into your own words. Instead of taking detailed notes, listen for several minutes without writing anything down, and then write a brief summary of what was said or presented. Remain vigilant for anything of value in what the speaker is saying. Imagine yourself on a treasure hunt.

Another cure for boredom is to become a critic. Look for anything the person is saying that you may disagree with. Tear apart what the speaker is proposing. Challenge the topic constructively. Weigh what's being said against your own belief system. Write down ideas on how the speaker could have made the presentation more interesting. This is called *active listening*. By looking for something to critique, you focus your mind on taking in, processing, and fully understanding what's being said so that you can offer an effective rebuttal. It keeps your mind actively engaged and prevents it from wandering off.

4. FORGETFULNESS

Have you ever been introduced to someone and immediately forgotten their name? Have you ever observed someone talking to you without remembering much of what they said? It's not that we don't hear what the other person is saying; it's that we don't care, or we're so focused on what we're going to say in response that we put little effort into processing or remembering what the other person is telling us. As a result, we engage in conversations that are a complete waste of time—for us and for the others involved.

I have two suggestions for overcoming this type of forgetfulness or, more accurately, lack of concentration:

- Work on your self-motivation. This is a listening skill. Create the frame of mind to listen by practicing listening. Create your own listening exercises at work, on the road, at parties, and with podcasts or other forms of audio information. Test yourself on what was said. Force yourself to concentrate, and your listening will improve.
- Concentrate on key selection: linking key words with images. Remember that memory isn't linear. Your mind doesn't remember lists, lines, or complete sentences in much detail. It works better by connecting words to mental images and concepts. As you listen to the person or engage in conversation, build a linked pattern in your mind—a visual map of ideas.

By practicing these techniques, you'll improve your ability not only to recall information but to understand what you're hearing and ask intelligent questions. You'll notice improvement both in business and personal interactions.

5. INDISTINCT AUDIO

Indistinct audio mostly involves people who talk softly or fail to properly enunciate when they speak. If you're ever unable to understand what someone is saying, you owe it to yourself and that person to say so. Simply say something along the lines of, "Excuse me, I'm having trouble hearing what you're saying."

Auditory processing disorder (APD)

The problem isn't always with the speaker. There is a medical disorder that can impair an individual's ability to distinguish between sounds. People with auditory processing disorder struggle to distinguish between subtle differences in sounds. Your friend suggests, "Let's go for coffee," and you hear, "Let's go cough." Or someone says, "I like your hair," and you hear, "I don't care."

You might describe this as "creative hearing," but it's actually a medical condition that affects about 3–5 percent of grade school children, who are unable to understand what they hear in the same way other children do. It's not that their hearing apparatus is damaged but that their brain misinterprets the audio input, especially speech. Children with APD have no trouble passing their hearing tests, but they often struggle with following directions and communicating effectively.

Likewise, as people get older, they may struggle more with auditory processing than actual hearing, and it's important to distinguish between the two. Giving a hearing aid to someone with a processing disorder probably won't do much good. What can help is removing distractions and being able to see the person who's speaking, so the person's lip movement can provide confirmation of what he or she is saying.

In most cases, speakers want you to hear and understand what they're saying, so they'll adjust accordingly, speaking louder or more clearly. They'll be glad you pointed it out to them.

The 20 Keys to Effective Listening

Over the course of my career assisting people in all walks of life improve their listening ability, memory, and recall, I have identified 20 keys to effective listening. Practice as many of these as possible. The more you can put into practice, the more skilled you will be at gathering and processing oral input. Here are the 20 keys to effective listening:

1. Take care of your hearing apparatus, which is much more sophisticated and delicate than you may think. If anyone questions your ability to hear, visit an ear, nose, and throat (ENT) doctor or an audiologist to have your hearing checked.
2. Train yourself to listen analytically. Practice listening to a cacophony of sounds and shifting your attention from one to another.
3. Maintain general physical health. Sound mind, sound body. When you're physically fit, especially aerobically fit, all your senses improve, especially your hearing.
4. Listen opportunistically. Even when you're bored, ask yourself, "What's in it for me?" In other words, be positively selfish. Insist that you're going to get something out of whatever's being said. Often we receive the most value in areas where we least expect to get something of value (assuming that we're paying attention).

5. Listen more than you speak. Shakespeare said, "Give every man thine ear but few thy voice," meaning, listen far more than you talk. Try to withhold judgment until you understand what the speaker is saying and you have gained something from it. Allow yourself to hear the whole story before you react to it.
6. Listen optimistically. If you go in with the idea that you have a chance to gain something, you greatly enhance the chance that you will. And if you have the right attitude, the whole experience will be considerably more pleasant.
7. Challenge your brain. You can stimulate your brain and improve your learning capacity if you expose it from time to time to more advanced material than you're accustomed to. Exposing your brain to challenging material may be difficult, but it maintains your enthusiasm and your ability to listen and learn.
8. Consciously put forth an effort to listen. Just decide that this is a part of your information processing arsenal and that you will be an active listener. You won't just fake it; you will be active.
9. Use synesthesia—the mental ability to blend your amazing senses. When you listen, be sure to engage, to lock in your other senses, especially sight. The more you link words to other words and sensations, the better your hearing, the better your attention, the better your understanding, and the better your general learning.
10. Keep an open mind. Don't let your emotions run away with you. Remain objective regardless of what the speaker says. In other words, try to interpret his or her words in a positive light even if you disagree.

11. Use your natural brain speed. You can listen four to ten times faster than the average person speaks. Because of that, we tend to daydream. Instead, use your brain speed to anticipate, organize, summarize, weigh, compare, and interpret body language. This will all help you concentrate.
12. Judge the content, not the delivery. The speaker's size, weight, stance, accent, clothing, and hairstyle are not important to the topic. Listen to the message.
13. Listen for the ideas. Your brain works better when it can grasp the whole rather than simply the parts, so listen for the central themes rather than individual facts. A fundamental understanding of the main topic provides a framework for organizing and understanding the details.
14. Instead of jotting down a list of notes, create a Mind Map. Draw an image at the center of it to represent the main idea, and then place your key recall words branching out from that center image. By using a Mind Map, you will dramatically increase your comprehension, understanding, retention, and recall. Mind Mapping engages your left and right brain. And as you know, that will improve your overall listening and recall performance.
15. Disregard distractions. Notice that I didn't say *avoid* them, because that's not always possible. Instead, remind yourself that you can consciously block them out. Your brain can filter out anything it doesn't want you to pay attention to if you allow it to.
16. Take breaks when possible (not when the speaker is talking, obviously). Give your brain time to integrate

everything and to create more opportunities for primacy and recency, as discussed in chapter 4.

17. Use your imagination. You may think that listening is a left-brain activity because it deals with words, but you can make it a whole-brain activity by creating mental images of the words and ideas you're receiving.
18. Listen with active posture. Your old grade school teacher probably scolded you to sit up because she knew that slouching detracts from your listening abilities (and she was right). You needn't be rigidly upright, just poised and alert.
19. Know that you can improve with age. They idea that memory deteriorates with age is a myth. Your listening skills and memory will improve if they're given the proper mental environment and you're practicing these 20 principles for good listening.
20. Practice speaking. This gives you the perspective from the other side of the podium. Knowing what you know about how people listen best, work on your speaking skills and make yourself an all-round communicator. Those who speak well listen better too.

Note Taking

Good listening skills can increase your performance at school and work and help you build stronger relationships with better understanding, so just imagine how much better you would do if you had a system for taking notes that could help you remember everything. Some people take notes only to discover later, when referring back to them, that they have no idea what they are referring to. It's almost as though they weren't even present at the meeting or presentation.

WORDS AND THEIR HOOKS

To improve your note taking skills, a basic understanding of the nature of words can help tremendously. Processing information, understanding the way in which your brain takes in the billion bits of data it receives daily, is arguably one of the most important things you'll ever discover about your brain.

Every word, every color, every number, every sound, is *multiordinate*. That is, each word is like a little central ball surrounded by a plethora of hooks. These hooks enable the word to attach to and associate itself with other words, images, and sensory data, to give the word and whatever it's attached to different meanings and significance.

Think about the word *run*, for example. One hook from *run* can be to run the four-minute mile. Another can be to run a printing press. Still another meaning can be found when you say she has a run in her stocking or you have run out of money. In the same vein, each brain is different from every other brain. Even if two people are enjoying the same experience at the same time, they are in very different universes, which means that each person will have an association for any given word, any given piece of data, that is different from every other person's.

For example, let's take another word: *leaf.* There will be a different series of images for each person who reads or hears it. For one person, the word may conjure up a mental image of a tree covered with green leaves. Another person may imagine a pile of autumn leaves, all brown, red, and yellow. To a third person, the word may remind her that she needs to rake her leaves or clean the leaves out of her gutters. Someone else may have fallen from a tree and associate leaves with pain.

Each one of us has our own associations. The mind picks the hooks that are the most obvious or sensational or those that evoke the most vivid images or strongest emotions. From that hook, the mind will be led down a path that is far more creative than mere recall.

Unfortunately, the mind, left to its own devices, will construct a story that's interesting but may not be particularly helpful for remembering. So when you're taking notes, use key recall words instead of key creative words. For example, suppose you're listening to a musical composition and want to describe it using words that will help you recall the composition. A word like *bizarre* is not specific enough to be a key recall word. It doesn't bring a specific image to mind, which is what you need. The word *Baroque*, on the other hand, would be a more effective recall word because it hooks the musical composition to other pieces of Baroque music, art, and architecture.

Effective note taking relies on very descriptive key recall words. A single key word or phrase should bring back an entire range of experience and sensation. There's no need to write out complete sentences to remember what you heard or saw. In fact, it's counterproductive. It's more effective to create memorable links and hooks.

About 90 percent of the words normally used in note taking are unnecessary for recall. That means you waste time writing them, and in the process, you're shifting your focus away from what you're reading or hearing. When you return to your notes later, you waste time rereading those unnecessary words and sifting through a lot of unnecessary words to find your key recall words.

Moreover, the connections between the key words are interrupted by words that separate them, and because memory works by association, the unnecessary words weaken the connection between the key words, thereby impairing memory and recall. And the fact that the unnecessary words separate the key words means you're going to waste more time getting from one key recall word to the next, which means there's less chance the connection will be made.

By grasping the key word concept, you are taking a major step in your ability to easily process information, and you'll also learn how to receive information at absolutely incredible rates of speed.

EXERCISE 1: NOTE TAKING WITH KEY RECALL WORDS

In his best-selling book *Thriving on Chaos*, Tom Peters, a leading business consultant and author, has this to say about listening: "Become a compulsive listener. Today's successful leaders will work diligently to engage others in their cause. Oddly enough, the best way by far to engage others is by listening, seriously listening to them. If talking and giving orders was the administrative model of the last 50 years, listening to lots of people near the action is the model of the 90s and beyond."

For this exercise, you need to recruit someone to play the role of your manager at work, who's giving you verbal instructions on how to conduct workshops. Your manager will read aloud the two short paragraphs of text that follow this paragraph. Do *not* read the paragraphs to yourself. As you listen, jot down key recall notes in the space provided following those two paragraphs. If you can combine them into Mind Map form, great. If not, don't worry; you'll work on Mind Mapping in more detail in the next chapter. Listen only once, just as if your manager were instructing you. Remember that the central idea is the workshop, so make notes around that central idea.

When your "manager" is finished reading and you've completed your notes, use your notes to answer the questions that follow. Ready? Here are the two paragraphs that your "manager" needs to read to you:

> You both will be conducting our next workshop on March 18, I think, about our corporate disability benefits. Since the participants will have to complete a few exercises and since they have to take your message back to their departments and be able to answer any questions about the benefits, it's important that you establish a good learning climate for them. The first few minutes of any workshop are really important, so you'll need to make the introductory comments interesting and relevant and take no more than seven minutes. And since there are 53 people in the workshop, one of you, whoever's not presenting, should be checking the body language of the group. You'll be able to tell if they're with you or not.
>
> I want each participant to be personally greeted by you. People react well to a handshake and a smile. Be sure the room is comfortable and everyone knows where the various facilities are. Be sure you announce the times for the

break and lunch. I have to warn you, many of people who will be attending don't want to be there. In fact, I heard the marketing department almost had hand-to-hand combat over which supervisor had to attend. Look, it's 10:30 now, and I have to meet with Ted Koch about our fiscal year budget. We need to meet again to review our objectives for the meeting. Can you both meet me here at, say, 2:00 this afternoon?

Your Notes

EXERCISE 2: ANSWERING QUESTIONS BASED ON WHAT YOU HEARD

Using the notes you took when listening to your "manager," answer the following questions:

1. What is the subject matter of the workshop?

   ____

2. What are the workshop participants expected to do when they return to their departments?

   ____

3. How many people will be in the workshop?

   ____

4. How should the participants be greeted?

   ____

5. How long should the introduction be?

   ____

6. Which department had a fight over who is to attend?

   ____

7. When does the manager want to complete the discussion?

   ____

My guess is that you answered most of the questions correctly. If you didn't, don't despair; you'll improve with practice. But give that listening quiz to someone who hasn't taken key word notes and see how well they do. You'll impress yourself with the difference.

Check your notes. If you're still not sure about answers, have your "manager" read aloud the two paragraphs again while you add to your notes.

EXERCISE 3: MIND MAP OF NOTES FROM 20 KEYS TO EFFECTIVE LISTENING

Based on the notes you took, create a Mind Map to help you remember and recall the workshop instructions and guidance that your "manager" provided. Or return to this exercise after completing chapter 9, where I provide additional guidance on creating Mind Maps.

Speed-Reading

Most people intuitively know that they can read better. Some say they have a vision problem. Others struggle with speed, comprehension, memory, or recall. Some are paralyzed by the thought of reading, especially from a big book or report. Some people are put off by an author's style. Some can't handle the vocabulary. None of these issues is a monumental problem to overcome.

The source of most people's struggles with reading can be traced back to how they were taught to read. Over the past couple of decades, we have made great strides in understanding how the brain functions, but virtually none of it has been incorporated into our methods of early learning. Our children are still being taught the way we were, the way our grandparents were, the way we've been taught for centuries—the see-and-say method, for lack of a better term.

Another prominent method is phonics. You learn the sound of each letter in the alphabet, then blend the letters into syllables and then increasingly difficult words, for example, "c-a-t," "cat." Actually there's nothing wrong with these methods. It's just they don't teach the entire reading process. They don't teach speed. They don't teach retention. They don't teach recall. They don't teach selection, rejection, note taking, concentration, appreciation, analysis, organization, motivation, effective type styles, or dozens of other factors that affect the overall reading process, that indeed *are* the reading process.

You've already learned how the brain approaches the processes I've just mentioned. But one that we haven't dealt with

Speed-reading apps

Increasingly, people are reading everything from blog posts to entire books on their smartphones, so it should be no surprise that the focus of speed-reading programs is shifting to smartphones as well. Multiple speed-reading apps are now available for smartphones, including the following:

- **Speed Read** includes a play button that you tap to launch an indicator that moves across the text to guide your reading. You tap the + or – button to increase or decrease the indicator's speed. Additional features enable you to bookmark and highlight text and even have the book read to you orally via text- to-speech.
- **Reedy** enables you to read books and web files in multiple file formats, adjust the speed, move forward or back, and check the time left to complete your reading based on your speed and the remaining words. By default, it slows the reading speed around complex words and punctuation.
- **Quickify** enables you to read content from web pages and most file formats, bookmark content, and have text read to you orally via text-to-speech. It also organizes all your content in a library for easy access.

yet is the speed of reading. You really can go a lot faster than you currently do. You really can.

UNDERSTANDING EYE MOVEMENT IN READING

What often slows down the reading process is the lack of fluidity in eye movement. When asked to show with their forefingers the movement and speed of their eyes as they read, most people move their fingers along in smooth lines from left to right, with a quick jump from the end of one line back to the beginning of the next, as shown in Figure 8-1, and they think they spend about a quarter to a full second per line.

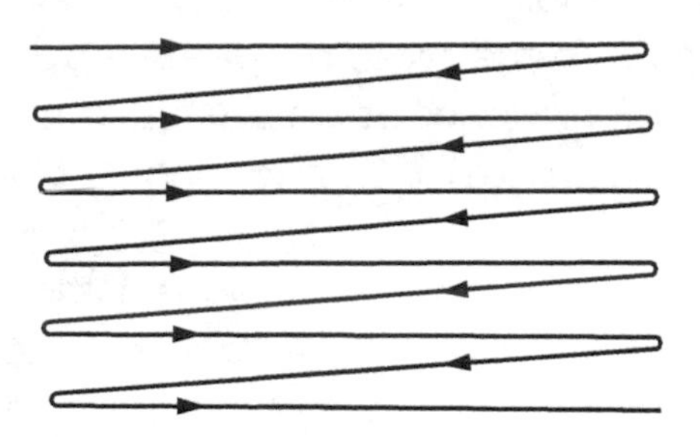

Figure 8-1: How people with no knowledge of eye movement think that their eyes move as they read a column of text. Each line is thought to be covered in less than one second.

But they're wrong. If you could read one line per second of standard print in a standard-sized book, you would be reading at an effective speed of about 600 to 700 words per minute (wpm), but the average person reads at about 250 wpm. So people think they're reading faster than they really are.

Here's another fallacy. If people really move their eyes that smoothly over a page of print, they wouldn't be reading. The eye can see clearly only when it's stopped. If you look at Figure 8-2, you'll see how the eyes really work. Instead of moving in smooth lines, they make a number of fixations on each line of print. They must, or they can't see. Slow readers make many more fixations than fast readers.

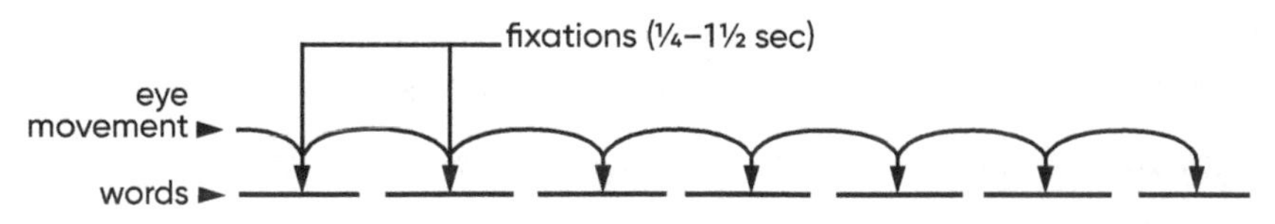

Figure 8-2: Diagram representing the stop-and-start movement of the eyes during the reading process.

The jumps themselves are so quick as to take almost no time, but the fixations on single words or phrases can take anywhere from a quarter to one and a half seconds. People who normally read one word at a time—and who skip back over words and letters, as shown in Figure 8-2—are forced by the simple mathematics of eye movement into reading speeds that are often well below 100 wpm. This means they will not be able to understand much of what they read or to read much.

THE THREE SECRETS TO READING FASTER WITH BETTER COMPREHENSION AND RETENTION

You can train yourself to read faster with better comprehension by practicing the three techniques described in the following sections.

Read to gain a general understanding of the topic

One of the problems slow readers have is that they go back over words and phrases they don't understand. Ninety percent of this regression is due to a desire to fully apprehend the meanings of the words and phrases, which isn't necessary for a general understanding of the topic. The other ten percent, the words or material you really don't understand, can be noted in Mind Map form to be checked later.

Thinking ahead and guessing the meaning is perfectly acceptable. In fact, it makes you a more active reader. You're challenging yourself to pull meaning out of context, which supports your brain's ability to create linkages for better retention and recall.

Reduce the time you spend on each fixation

Fixation on specific words runs between a quarter of a second to one and a half seconds. You don't need to spend so much time on a word. Try to push yourself towards a fixation of around a quarter to a half a second per word.

While that may seem too fast at first, it really isn't. Tests have proven that the eye can take in as many as five words in one one-hundredth of a second. A slow reader will fixate as many as 500 times per page. That's very slow and really holds you back. A faster, better reader will do the same page at maybe 100 fixations. Assuming that both readers spend a half second per fixation, the faster reader is 200 seconds faster per page—that's three minutes and 20 seconds faster per page! If both were reading the same 300-page novel, the faster reader would save 1,000 minutes. That's over 16 hours faster!

It follows that the faster reader conserves energy and doesn't tire as quickly as the slower reader. Fewer fixations results in much more efficient use of the eye muscles, resulting in less strain. And less strain makes reading physically easier. You get all the advantages.

Take in more words per fixation

The third way to dramatically improve your reading speed is to take in as many as three to five words at a single fixation. The mind can deal equally well with groups of words as it can with single words. When we read a sentence, we don't read it for the meaning of each individual word. We read it for the meaning of phrases in which the words are contained.

USING YOUR FINGER TO TRAIN YOURSELF TO READ FASTER

Your grade school teacher probably instructed you *not* to use your finger to point at words in a book while you're reading because it slows you down. We now know that your grade school teacher was wrong. Pointing, when done properly, doesn't slow the eye down. What your teacher should have told you was to move your finger faster along the print. A finger, pen, pencil, or any sort of visual aid, helps to guide the eye. Studies show that eye movement is much more efficient when you use something to guide it.

To practice reading faster, trace your forefinger along each line of print. Since we imagine that we read faster than we actually do, this process makes it feel as though you're going slower, but you're not.

Second, practice looking at more than one line at a time, expand your range of vision, and try different patterns with your finger or pen—diagonal, curving, or straight down the middle of the page. See what works best for you, practicing at first only with light reading. You don't want to practice with an engineer's instrumentation handbook.

GIVING YOURSELF A TIME LIMIT

Another technique to increase your reading speed is to give yourself a 20-seconds-a-page limit. In that 20 seconds, try to take in as many words from the page as you can. Go for about seven or ten pages at that pace—for example, one chapter or a group of pages. When you've finished, write down as many words as you can remember from that high-speed reading.

At this stage, don't worry about comprehension; focus on basic word recognition. Then look at how your brain is linking those words. Using the related words, what kind of story line has your brain come up with? What integral meaning is your brain drawing from those words

Now go back and read the same pages at 40 seconds a page for better comprehension. You'll be amazed at how much you retain and comprehend at those high speeds.

This kind of exercise is very similar to physical training. The more you exercise, flex, and stretch, the greater your endurance, strength, and flexibility. It is the same for your brain, so push yourself appropriately. Condition your mind to experience higher speeds, and it won't take long before you realize that you are reading at least twice as fast as you were, simply by engaging in accelerated reading. Your brain thrives on it.

By now all of your senses should be operating on a new plateau. You've learned a lot about yourself and your brain power. You've probably started to use your new skills to make exciting new gains in your work and personal life. Now you're about to enter the final stage of this program: putting the entire process together to feel the superpower of total information management.

CHAPTER 9

Thinking in Pictures with Mind Maps

The human brain is an incredible pattern-matching machine.

—JEFF BEZOS

As you're about to put this entire process together, let's review what you've accomplished so far. You've learned that no matter what you've already achieved, you haven't even come close to using your phenomenal brain power, and that there's no excuse for not doing better. Genetics may determine your upper limits, but they do not get in the way of significantly improving beyond where you are today.

I have introduced you to the most up-to-date information on human processing skills, including logic, numeracy, analysis, order, sequencing, and imagination. You've learned how to change habits for the better, how to use the rest-activity cycle to improve brain power, how to face change, and how to organize your life into chunks. You've learned that if you say you have a terrible memory, it's because you don't care to do anything about it. You know that your brain has the capacity to hold billions and billions of pieces of information and that you can recall them simply by developing a retrieval system that takes advantage of your brain's latent ability.

You've had a chance to improve your self-organization, understand how you can increase your life expectancy, create lists, and develop an image of your future. You've become your own life guide by dividing your life into seven or fewer categories and then chunking them to gain control over yourself and all you do. You've thought, maybe for the first time in a long time, about your need for love and friendship, self-development, and financial and business success. You've had a chance to integrate improvements in listening, speed-reading, and note taking, and you've been introduced to Mind Mapping, something that will be added to your thinking and organizational arsenal in this final chapter.

Exercise 1:
Preparing Notes for a Book You've Read

Think of about the last book you read. Now imagine that in a year's time, you will be required to lead a seminar on that book. Spend some time preparing notes about that book in the space provided below. Your notes need to be clear and detailed enough so that when you look them over a year from now, you'll have all the information you need to prepare for and deliver your seminar.

Seminar Notes

I've done this exercise in many of the courses I've taught over the years and found that about 70 percent of my students make notes in paragraph or sentence summary form. The other 30 percent use key phrases or key words, usually in some list form. Based on what you learned so far, you already know that there's a better way, and you're going to become an even better note taker as you work through this chapter. This is the better way.

The Limitations of Traditional Note Taking

Dr. Gordon Howe of the University of Exeter, along with other researchers, have done extensive studies to identify the type of note taking that's best for learning and recall. The seven types, from worst to best, are as follows:

1. Taking no notes at all
2. A complete transcript of the presenter's notes
3. A complete transcript of the listener's notes
4. Summary sentence notes provided by the presenter
5. Summary sentence notes created by the listener
6. Key word notes provided by the presenter.
7. Key word notes created by the listener (fewer key words, within reason, are usually better)

Key words are the most succinct and effective expressions for triggering immediate recall. One of my students described them as the stepping stones you use to cross a swamp—the swamp being the dense collection of paragraphs and sentences that contain the key words, but which in themselves aren't essential for recall.

The study didn't consider *shorthand*—a formal system of rapid handwriting that uses abbreviations and symbols to represent words, phrases, and letters. If you know shorthand, you're probably thinking that there's nothing wrong with writing everything out as long as you can write as fast as the presenter can speak, but that's what I call an "intelligence trap." You're doing something that you've been conditioned to do and has been successful in the past, assuming that it's the best way when it's really not. You've just become very comfortable with it.

Let me prove it to you. I'd like you to think about all the sentences of 10 words or more that you can recall ever having spoken, read, or heard during your lifetime. Now forget any sentences you've made a special effort to remember through repetition, such as lines from a poem, a play, or a song. Take your time. I doubt that you can recall more than one or two 10-word sentences, if any.

In all the years I've been delivering my brain power seminars, nobody has come up with more than ten 10-word sentences that they can recall verbatim, even though they probably have billions stored in their memory banks. One student who worked in advertising recalled a compliment she received from her boss six months after starting her new job. He called her into his office and said, "Young lady, I'm very pleased with the job you're doing, and I believe you have a fine career ahead of you."

That's an example of outstandingness. We tend to remember and recall what we find most outstanding. That was a major moment in her career, and her memory of it was crystal clear. When it happened, she probably played it over and over in her mind. However, that was the only sentence of at least 10 words that she could recall.

The reason we can't recall more sentences is that we don't think in sentences, even though we've heard, read, spoken, and written millions of sentences. We recall key words and key images. Even when you're delivering oral or written instructions, you're usually thinking in key words and images first, then constructing sentences to describe what you're thinking.

Optimizing Your Note-Taking Efficiency

Only about 10 percent of what people write is of value (that's before they learn my system, of course), which means that doing it right improves your note-taking speed by a whopping 10 times at a minimum. The icing on the cake is that by adding a few doodles to the notes, you'll dramatically improve your recall.

What helps us remember the most are linkages and outstandingness.

You can make your key recall notes incredibly effective by using arrows, colors, special codes, and tiny drawings. Make connections between key words that immediately show you where the relationships lie. Once you've made these linkages, you can make them outstanding by adding color or size variations or making them look three-dimensional. You can outline them, underline them, do anything you want to make them stand out.

If you know shorthand, you can even integrate it into your Mind Maps. The key is to simply eliminate unnecessary and irrelevant words. Once you're good at confining yourself to key words, you can save even more time by writing your key words without vowels—using only consonants. Fr xmple Im sr yll hv no dffclty rdng ths bbrevtn. It is surprising how readily your brain fills in the gaps. Your brain won't give up until it makes sense of the text.

Improve Your Note-Taking Speed

One of the problems with traditional note taking is that it draws your attention from the presenter and the content being presented. Another problem is that people generally speak much faster than any note taker can write. The average person speaks approximately 100–130 words per minute, whereas the average person can write only about 25–35 five-letter words per minute.

You can dramatically improve on that average writing speed by giving yourself one-minute practice sessions—pushing yourself to write as many words as possible in 60 seconds. However, speed isn't the point here. You need to be able to relax. Don't get writer's cramp when taking notes. Don't hunch your shoulders, as many people do when taking notes. They tire themselves out. A relaxed upright posture will optimize your note-taking performance.

The best way to improve your note-taking speed is to take fewer notes. Write and draw only the most outstanding, descriptive, and impressive words and images.

Mind Maps: The Better Way

Mind Maps are tools for managing information using a combination of graphics and text. They enable you to organize information for enhanced recall and use that information to create new thoughts, ideas, images, and concepts. You might think of them as handmade infographics. In addition to helping you organize and recall information, Mind Maps provide you with all the elements you need to solve problems and to make sound decisions in your career, business, and personal life.

A Perfect Fit with the Brain's Inner Workings

The brain handles information much better when that information is designed to slot into all the multiple locks that the brain has. The brain is multidimensional in nature and will much more readily understand, appreciate, and recall information that's presented multidimensionally. Words, lines, arrows, colors, codes, images, and so on all contribute to making information multidimensional.

In order fully to utilize your brain's capacity, you need to consider each of the elements that add up to the whole, and you need to integrate those in a unified way. The answer is the Mind Map. The Mind Map does that for you.

Mind Maps are aligned with the way your brain functions. They can be used in every activity involving thought, recall, planning, or creativity. When many people see a Mind Map for the first time, they think it's too dimensional and chaotic to translate into linear presentations, such as speeches, talks, or articles. That is simply not true. Once a Mind Map is completed, all you need to do is to select the order in which to present the information, and that will translate it into a linear presentation.

In fact, students at the University of Oxford in England were found to complete their essays in one third of the previous time required after switching to Mind Maps. Likewise, managers and personnel in major corporations, including Digital, IBM, EDS, General Motors, and Hewlett-Packard, have experienced a similar boost in performance—one third of the time spent and a 300 percent increase in value when they used a Mind Map as opposed to their previous older methods. The Mind Map really does work. It's faster *and* better.

Mind Mapping Laws

Mind Maps are not only a way of recording and managing the mountains of information that you have to process every day of your life. They're also the key for helping you create new thoughts, ideas, images, and concepts from that information. Even though the Mind Map concept requires some unlearning of old linear habits, it represents the way you really think and absorb information. You do not think in sentences or in lists. You think along the lines of connections made by the brain.

Recent biochemical, physiological, and psychological research continues to confirm what I have been teaching for years: the brain is nonlinear and is only restrained by our emphasis on linear thinking. Mind Mapping will free your various creative abilities, give you more hours in every day, and unleash your talents for information management. Here are the principle laws of Mind Mapping:

1. *Start with a colored image in the center.* An image often is "worth a thousand words" and encourages creative thought while significantly increasing memory.
2. *Use images throughout your Mind Map.* As with the image in the center, they help your memory. They stimulate left-cortical and right-cortical activity. They use your entire brain. They focus your attention. They help your understanding. They increase your thinking power.
3. *Print all words.* It may take a little bit more time to print than to write, but you'll save more than that time when you reread the Mind Map. A printed map gives a much more immediate, photographic, and comprehensive feedback. Overall, printing saves time.
4. *Print words on lines, connecting each line to other lines.* This ensures that the Mind Map has a basic structure, and it facilitates linking and association in your brain.
5. *Treat each word as a unit, one word per line.* This provides the most hooks per word and gives you maximum creative and memory flexibility.
6. *Use colors throughout the Mind Map.* They enhance memory, delight the eye, and stimulate the right-cortical process, while enabling you to think more clearly and effectively.
7. *Give your mind as much freedom as possible.* Suspend critical thinking. Any thinking about where things should go or whether they should be included will stifle creativity and slow down the process. This law applies specifically to brainstorming and creative thinking. Its

purpose is to free your mind, so let go. Fight the urge to be overly structured, rigid, or meticulous. The whole idea in brainstorming and creative thinking with a Mind Map is to let everything that crosses your mind flow out, no matter how ridiculous it is.

The idea is to recall everything your mind thinks of around the central idea. As your mind will generate ideas faster than you can write, there should be almost no pause; if you do pause, you will probably notice your pen or pencil dithering over the page. The moment you notice this, stop and carry on. Do not worry about order or organization, as this will in many cases take care of itself. If it does not, a final ordering can be completed at the end of the exercise. To a large extent, your brain naturally organizes information. Often the best course of action is to get out of the way. Don't overthink it.

Neatness

Many people worry about how neat their Mind Maps are, particularly since the final structure doesn't often become really apparent until the very end. If you're concerned about neatness, keep in mind that content is far more important than appearance. As a matter of fact, I often tell participants in my classes that notes that look messy are usually far neater in terms of their suitability to the brain. They show immediately the important concepts and connections. They give the real story of the content that is being summarized and distilled. You can revise or reconstruct your Mind Map in about

Mind Mapping software and apps

Personally, I prefer to create my Mind Maps by drawing them on a piece of paper. The physical activity of drawing the map provides kinesthetic reinforcement of the content. However, if you prefer working on a computer or smartphone and creating neater Mind Maps from scratch, check out the latest offerings in Mind Mapping software. You can find plenty of options for all operating systems, including Windows, Mac, and Android devices. Here are a few examples:

- **InfoRapid Knowledgebase Builder** is a knowledge management app that enables you to create links to articles, blog posts, websites, tweets, and documents and perform a full text search on all linked content. It can even generate Mind Maps from existing content, import tweets and Wikipedia articles, and export your knowledge base as a formatted HTML document, so that you can view it in a web browser.
- **GitMind** is available as an online application that you can use through any web browser. It's also available as a desktop application and smartphone app. It has several templates you can start with and a variety of shapes and colors for creating Mind Maps from scratch. GitMind also supports online collaboration for group projects.
- **miMind** is multipurpose, cross-platform, Mind Mapping software designed to create and share ideas and activities, such as project planning, brainstorming, designing, summarizing ideas and discussions, and doing project demos. The base version is free, but you can upgrade for more advanced features.

10 minutes anyway—and in fact, you should. The process of redrawing your Mind Map reinforces learning.

You're being asked to break loose from the chains that have bound you to the rigid structure you probably learned in school, which was reinforced in the work world. You should now be prepared to use your information management skills to become the person you always knew was just under the surface. You're now prepared to more fully understand that knowledge is power and that you have plenty of both.

Exercise 2: Mind Map about Yourself

The next page is blank with an outline of a human being in the center. The image is intended to be generic, but you can personalize it however you like: draw a face on it, add hair, color it, whatever. The image represents the central theme in a Mind Map about you: how you see yourself now, the way you want to see yourself in the future, or both. You can incorporate both into your Mind Map. Maybe you'll have one extension for goals, which can branch into several different areas, another for career, and a third for relationships.

Do you remember the exercise you did on the seven major divisions of your life? Perhaps you can branch out into goals and ideals from each of them. You are multifaceted. Don't waste time worrying about structure. Just let yourself go. Take 15–30 minutes to Mind Map yourself.

Well, how did you do? Was the exercise illuminating? Did you discover anything about yourself you weren't aware of? Through this exercise, some students discover that they're not very good at drawing straight lines, but that's not really

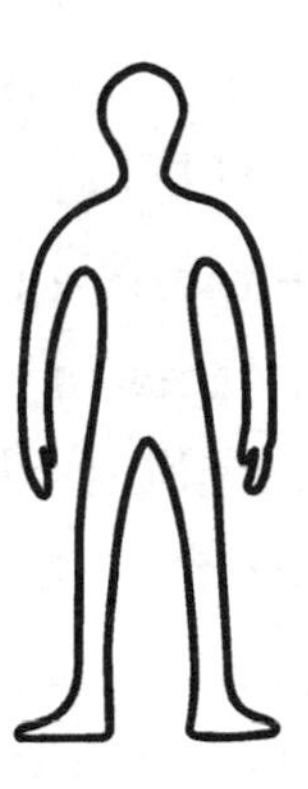

important. Drawing straight lines is not one of the keys to success. What's important is that you discover something about yourself you weren't already aware of—something that helps you function better at school, work, or in relationships or brings you closer to self-fulfillment.

What struck one student most was that the sense of time kept popping up, as in "I don't have enough time for exercise. I don't have enough time for sailing. I don't have enough time for reading." It became obvious that she needed to restructure her time commitments to do more of what she wanted to do.

Discovery is a valuable side benefit of Mind Mapping. The more you branch out, the more you spark ideas and images. Each word evokes lots of images in your brain. All you need to do is create a structure that allows them to flow. One idea can spark another. You may also find that something appears over and over again, bringing to a conscious level what may have been lurking in your subconscious. Creating a Mind Map about yourself can become a highly personal brainstorming session.

It can also be effective in a group. For example, if you're starting a business or a nonprofit organization, Mind Mapping can help you define its mission, vision, and values.

Mind Mapping for Lectures and Meetings

When you're taking notes, say at a meeting or a lecture, I recommend that you take them within a Mind Map structure. Use the left-hand page for mapped information and the right-hand page for more linear or graphic information—things such as charts or graphs, formulas, and special lists. When you're finished, you can combine them into a single, comprehensive Mind Map that will give you a more complete and coherent picture of the material than any notes you've taken previously.

In the following sections, I provide more specific guidance for using Mind Maps in lectures and meetings.

LECTURES

When taking notes during a lecture, I recommend using a large blank page to enable your brain to see the big picture. Keep in mind that key words and images are essentially all you need. Also remember that the final structure will not become apparent till the end. Any notes you make will probably be semifinal rather than final copy. The first few words noted may be fairly disconnected until the theme of the lecture becomes apparent.

Understand the value of messy as opposed to neat notes. Neat notes are traditionally organized in an orderly and linear manner. Messy notes are "untidy" and all over the page. Many people feel apprehension at having a scrawled, arrowed, nonlinear page of notes developing in front of them. But the word "messy" here refers to the appearance and not to the content.

In note taking, it is primarily the content and not the appearance that is of importance. Notes that appear neat are messy in respect to information. The key information is disguised, disconnected, and cluttered with many irrelevant words. Notes that appear messy are informationally far neater. They immediately show the important concepts, the connections, and even in some cases the crossings-out and the objections.

Mind Mapped notes in their final form are usually neat in any case, and it seldom takes more than ten minutes to finalize an hour's note on a fresh sheet of paper. Reconstructing a "messy" Mind Map is a productive exercise, particularly if the learning period has been organized properly.

MEETINGS

Meetings, notably those for planning or problem solving, often degenerate into situations where each person listens to the others only in order to make his own point as soon as the previous speaker has finished. In many cases, the individual or group requesting the meeting has an agenda and a clear idea of the solution to the problem. They're simply gathering everyone together to create a false sense of consensus—to make everyone who attends the meeting feel as though they had an opportunity to provide some input when they really don't.

These problems can be eliminated if the person who organizes the meeting uses a Mind Map structure by using the following approach:

1. At least one or two days prior to the meeting, contact the participants to let them know what the central theme of the meeting will be and to come prepared with their ideas. Meetings are much more productive when participants are prepared to contribute.
2. On a board or easel at the front of the room, draw a picture at the center to illustrate the central theme of the meeting.
3. (Optional) Branch off from the central theme to include any subthemes you want to cover.

Avoiding the black hole of business meetings

One of my students described problem-solving meetings at her company as "a black hole in space that swallows up each speaker. Everybody wants to make his or her point, period. If someone listens to you at all, it's because he wants to demolish your key points."

Truth and the best ideas are often sacrificed at the altar of ego, and a lot of time is wasted. The solution that's approved isn't always the best; it's simply the one that's jammed down everyone's throat by the loudest or highest-ranking members.

4. Open the discussion and add each participant's key points to the Mind Map, connecting them to the central theme or one of the subthemes. Consider asking each participant where their key point should be added or (better) have participants come up to the board and add their ideas to the Mind Map themselves.

These are the advantages of incorporating Mind Mapping into meetings:

- The contribution of each person is recorded, not distorted.
- No information is lost or ignored.
- Ideas gain priority over the status or personality of the person proposing the idea.
- People talk more to the point, thereby eliminating digressions and long, meandering expression. Meetings proceed at a faster pace and are consequently shorter.
- After the meeting, each participant will have a mapped record of what was said. The information won't be lost

by the next morning. Tip: Encourage participants to take a photo of the completed Mind Map with their smartphones.

- Everyone becomes actively involved in the growth and development of the Mind Map rather than being concerned solely with taking notes and gathering information. This increased engagement leads to greater critical and analytical success and considerably more understanding, not to mention improved memory and recall.

Mind Maps are external "photographs" of the complex interrelationships of your thought at any given time. They enable your brain to see itself more clearly and greatly enhance the full range of your thinking skills: they will add competence, enjoyment, elegance, and fun to your life.

Mind Mapping and Note Taking While Studying

Mind Mapping and note taking aren't limited to lectures, presentations, and meetings. As you read and study any subject, whether for a course you're taking at school or something you're interested in learning about on your own, Mind Mapping, along with marking up the texts you're reading, can tremendously help improve memory, recall, and overall understanding. Take the following two-pronged approach:

1. Mark up any and all texts you're reading as part of your study program—books, reports, white papers, and so on.
2. Create and maintain a growing Mind Map.

Mark Up Your Texts

Whether you're reading a book, a research report, a white paper, or any other text that's part of your study materials, mark it up. Your markup may include any or all of the following:

- Underlining or highlighting key recall words
- Personal thoughts elicited by the text
- Critical comments
- Marginal straight lines for important or noteworthy material
- Curved or wavy marginal lines to indicate unclear or difficult material
- Question marks for areas that you wish to question or find questionable
- Exclamation marks for outstanding items
- Your own symbol code for items and areas that relate to your own specific and general objectives
- Mini Mind Maps in the margins

If the book is not valuable, marking can be made in color codes. If the book is a cherished volume, markings can be made with a very soft pencil. If the pencil is soft enough and a very soft eraser is used, the damage to the book will be less than that caused by the finger and thumb as they turn a page.

Mind Mapping as You Read

Just as you can create Mind Maps in a meeting or when listening to a lecture or presentation, you can create a Mind Map as you read. During your initial encounter with a text, as you are browsing through the text, draw a small image in the center of a blank page to represent what you think is the central theme of the text, and then draw branches out from that central theme to highlight what you think are important subthemes. Simply by reading the title and subtitle of the book and skimming through its table of contents and index, you should be able to develop a pretty clear idea of what the book is about—its central theme.

As you dig into the book, add details to your Mind Map. In chapter 7, I introduced you to the seven-step process that superlearners follow whenever they encounter new subject matter. By starting your Mind Map with a central image representing the central theme, you've already taken the first step. Now you're ready to flesh out your Mind Map by asking questions, skimming material throughout the text, and conducting your inview and review. Remember, inview involves reading what's most relevant to you and what you think is most important and skipping anything you already know or that's irrelevant or unimportant. When you conduct your review, you're rereading anything you're unclear about or that you overlooked.

Commit to Your Future Development

As you come to the end of this book, there are a number of ways in which you can continue to practice and become more aware of as part of your future development. Here's a rundown of key points and actionable items covered in this book:

- As you study, take frequent breaks to maximize opportunities for primacy and recency and to provide your brain with time and space to integrate the new learning into your existing understanding. Taking frequent breaks is one of the best ways to increase retention and recall. You give the information time to settle in, sort itself out, and become more complete.

- When you feel you have a solid foundational understanding, review the material. Browse through it again, because when you hit that high peak of understanding and memory, that's when the entire entirety of what you learned clicks in. It'll be like watching your favorite old movie again and appreciating the nuances even more, catching details that you overlooked the first or even the second or third time you watched it.
- Continue to stimulate and use the entire range of your cortical skills. Whatever you are strong in now, continue to use and develop. Whatever areas are weak, continue to strengthen and grow. Yes, that includes mathematics. Now that you know that your brain is a natural, delightful, genius mathematician, develop those skills. Practice mental literacy with mathematics, and when you find people using calculators, play games with them. See if you can calculate the answers in your head before they do.
- Continue to develop your logical skills. The way in which you analyze and integrate information that comes in from all the sources around you is essential for your survival and success. Continue to develop and grow those skills. As you read or watch the news, remain skeptical. Challenge the sources of information. Look for bias in reporting. Question the logic behind the conclusions being drawn. Don't just passively accept what you're being told.

- Nurture your memory—the vast storehouse of information that is yours by right. Make optimal use of primacy, recency, and outstandliness to remember and recall key concepts and create vivid links between words and images. Remember the SMASHIN' SCOPE of memory:

  Synesthesia/sensuality — Number
  Movement and dimension — Symbolism
  Association — Color
  Sexuality — Order/sequence
  Humor — Positivity
  Imagination — Exaggeration

- Continue the process you started when you began to explore yourself. Continue to monitor who you are as a human being, your strengths and weaknesses, and how you can get the best advantage out of everything you do from this point on. That includes your finances. Get your cash flow, net worth, and financial planning in order.
- Start a diary or universal personal organizer (UPO) with the goal of creating and maintaining a growing book about yourself—a book that captures your life in words, images, and sensations to engage both your left-brain and right-brain thoughts and emotions. Don't simply log daily entries about what you did or what happened to you each day. Use your UPO to set goals, record your thoughts and plans, Mind Map your most creative ideas, and plan your time for optimal use.
- Within your diary or UPO, I strongly recommend that you now devise a plan for your ongoing education, your ongoing lifelong learning. Already begin to search out the

learning centers around you: places where you can establish new areas of interest, new hobbies, new activities. Plan that out. Begin the development.

- Engage and develop all your senses: sight, hearing, smell, taste, touch, movement. Take proper care of your sensory apparatus. Protect them from harm while providing your body with everything it needs to maintain optimal health. Shift your focus to different sensory stimuli throughout the day to sharpen your senses and reinforce neural networks in your brain that are connected to the senses.
- Develop your reading skills. You now know that you can read much faster with significantly improved comprehension, so continue to practice the speed-reading techniques in chapter 8. You should begin to eat books for breakfast—no longer one book every two years, but one book every couple of hours. Feed your brain that information. Remember the four foods: information, oxygen, good nutrition, and especially love and affection. Prioritize the care and feeding of your brain.
- To integrate all those developing skills, use your new major thinking tool: the Mind Map. Whatever thinking you want to do, as long as you need to externalize it, to put it outside, to make notes, use the Mind Map. Not only will it allow you to remember better, to create better, and to be more powerful in your general thinking, it will also give you artistic and aesthetic appreciation and will help develop all the mental skills you have. And remember that as you get older, that massive powerhouse, your brain, will get better and better if you use it. So use it, and use it well.

With the commitment you've made to this course, you're well on your way to putting what you've learned to work in your everyday life. You now know you have the ability to expand as an individual, develop yourself in ways you never thought possible, and manage information rather than letting it manage you.

At this point, I would normally wish you luck, but luck is not what you need. You simply need to put into practice what you learned from this book. By doing so, you will be in control of your own destiny. You will no longer be a victim of circumstances, bad luck, or misfortune. Your life will be a product of your knowledge and creativity, along with your courage to make bold choices and take bold actions. Instead of wishing you luck, I wish you a healthy, vibrant brain and a richer, fuller life.

About the Author

Tony Buzan (June 2, 1942–April 13, 2019) was an English author and education consultant who popularized the idea of mental literacy, radiant thinking, and a technique called Mind Mapping. His was the mind behind Mind Maps® and the concepts of Mental Literacy and Metapositive Thinking. He also founded the International Brain Clubs and was responsible for many developments in the theory and advancement of mnemonic systems and creativity.

Buzan was editor of the *International Journal of MENSA* (the high IQ Society) and also was the holder of the world's highest creativity IQ. Among the members of the Young President's Organization, he was affectionately known as "Mr. Brain."

He was a prolific writer, authoring more than 14 books (most on the brain, creativity, learning, and memory, along with one volume of poetry). His books, which include *Use Both Sides of Your Brain, Use Your Perfect, Memory, Make the Most of Your Mind, and Speed-Reading,* have been published in 50 countries and translated into 20 languages. *Use Both Sides of Your Brain* surpassed worldwide sales of 1 million and is a standard introductory text for staff training within IBM, General Motors, EDS, Fluor Daniel, and Digital Equipment Corporation, and for students of the Open University.

Tony Buzan was an international media star who featured in, presented, and coproduced many satellite broadcasts, television, video, and radio programs, both national and international, including the record-breaking *Use Your Head* series (BBC TV) and the *Open Mind* series (ITV); *The Enchanted Loom*, a one-hour feature documentary on the brain; numerous talk shows; and a video series called *Improving Mental Performance*—a three-part training package that introduced the major elements of his work to the international business community.

He was an advisor to governments and multinational organizations, including BP, Barclays International, Bell Telephone, AT&T, Rank Xerox, and Nabisco, and was a regular lecturer at leading international universities and schools.

Buzan was also an advisor to international Olympic coaches and athletes and to the British Olympic Rowing Squad as well as the British Olympic Chess Squads. He was a Fellow of the Institute of Training and Development, the Jamaican Institute of Management, and the Swedish Management Group, and was an elected member of the International Council of Psychologists. He was a member of the Institute of Directors, a Freeman of the City of London, and a Patron of the Young Entrepreneurs' Societies of both Bristol and Cambridge universities. Adding to his list of honors, including the YPO Leadership Award, was his recognition by Electronic Data Systems with the Eagle Catcher Award—given to those who attempt the impossible and achieve it!